Preface

This manual is produced by Environment Canada meteorologists with funding from Transport Canada through the Search and Rescue New Initiatives Fund and the Air Navigation System Development Fund. Its overall purpose is to contribute to flight safety in British Columbia and the Yukon.

This guide describes, in as much detail as is practical, local aviation weather hazards along the major VFR flying routes in British Columbia and the Yukon. The detailed local weather is presented graphically, using maps wherever possible and forms the final chapter. In order to make this manual self-contained, a number of initial chapters are also required to describe patterns and phenomena which affect the local hazards in Chapter 4. This Aviation Weather Hazards Manual commences with some basic meteorology followed by sections on specific hazards to aviation, local (small scale) weather and system (larger scale) weather.

When it comes to local, small scale aviation weather, experienced pilots are the best source of information. Much of the information in Chapter 4 was obtained through interviews with pilots throughout B.C. and the Yukon. This manual is written by aviation meteorologists who are not pilots and have little knowledge of other aspects of aviation. Thus, flying decisions such as where or when to fly, even if suggested by experienced pilots, will not be provided here. Instead, knowledge of local weather conditions will be communicated. For instance, suggestion that a particular option is drier than another does in no way imply a suggestion to fly a certain route.

The information presented in this manual is by no means exhaustive. The variability of local aviation weather in B.C. and the Yukon could result in a publication many times the size of this one. Thus, the authors have chosen to focus on VFR aviation along the common routes. By understanding some of the hazards along these routes, pilots may be able to relate the hazards to topography and weather systems in areas not covered. It is hoped that readers of this book will constitute itinerant pilots, pilots new to Western Canada and flying schools. In addition, experienced pilots may find it useful when flying to unfamiliar regions.

This publication was borne out of the successful marine weather hazards manuals which have been produced on both Pacific and Atlantic coasts. Coastal aviators in B.C. encounter many of the same hazards as their counterparts in the marine community. Many facts and details from the West Coast Marine Weather Hazards Manual are included. However, coastal pilots are strongly encouraged, if possible, to use this manual along with the enhanced detail included in the Marine Weather Hazards Manual.

A form is included at the end of the manual for provision of information, either to add or clarify facts provided. Any suggestions from pilots will be archived for use in future editions of the Aviation Weather Hazards Manual.

ACKNOWLEDGEMENTS

This manual was made possible through funding from the Search and Rescue Secretariat of Transport Canada as well as from the Air Navigation System Development Fund.

We would like to thank our partners at Transport Canada both nationally and regionally, specifically Pacific Region and Western Region. Special thanks also go to Owen Lange, marine meteorologist in Vancouver, who provided coastal details from the West Coast Marine Weather Manual.

Finally, the vast majority of the local weather information was provided by pilots throughout B.C. and the Yukon. Our job was made very simple due to the clear and concise information provided by over 100 pilots. Their openness and friendliness made our task very enjoyable.

Kent Johnson

John Mullock

July, 1996

Readers are invited to submit any additions, changes or comments to:

Aviation Weather Hazards Manual
Mountain Weather Services Office
Environment Canada
3140 College Way
Kelowna, B.C.
V1V 1V9
Canada

Aviation Weather Hazards of British Columbia and the Yukon

by
Kent Johnson
John Mullock

Graphics & Illustrations
Dave Rothwell
Source Graphics & Print Co. Ltd.

Production of this manual was funded, in part, by the National Search and Rescue Secretariat and the Transport Canada Air Navigation System Development Fund.

This manual was designed and printed by

Source Graphics & Print Co. Ltd.

Source Graphics and Print Co. Ltd.
1 - 1050 Leathead Road
Kelowna, B.C. V1X 2K1
(250) 765-6445
Fax (250) 765-7221
1-800-667-9266

Published by the authority of the Federal Minister of Environment

Catalogue number EN50-116/1996E
ISBN 0-662-244794-9

TABLE OF CONTENTS

Chapter 3

Chapter 4

Introduction

Meteorology is the science of the atmosphere. It embraces both "weather" and "climate" and is concerned with the state of the atmosphere as well as the complex interaction of the atmosphere and the underlying surface. Aviation takes place in the atmosphere and as such is inseparable from the "weather" side of meteorology.

It has been said that the mountains are the "mothers of climate." Nowhere else in Canada is such a statement so aptly demonstrated as in British Columbia and the Yukon. Shifting dramatically from the dripping, fog-shrouded rain forests of the West Coast through the parched valley bottoms of the Interior to the majestic snow-capped peaks and glaciers of the Yukon, few places in the world offer more visual splendors for a pilot and their passengers. At the same time, mountains also evoke rapidly changing weather conditions that all too often have contributed to a tragedy. Between 1976 and 1994, there were 419 flying accidents in British Columbia and the Yukon where weather has been identified as one of the contributing factors. In these accidents, 319 people were killed and 89 people injured seriously. Mountain flying itself is not inherently dangerous, rather it is the weather associated with these areas that tends to be unforgiving of the rash, the negligent and the unlucky.

This manual describes the different ways that storms approach the coast and how they are modified by mountainous terrain. While many specific localized aviation weather hazards are mentioned, it would be impossible to list them all, even if they were known to the authors of this book. The goal is to enable the aviator to recognize the processes which cause the specific hazards and hence be able to anticipate other areas which might experience the same conditions.

Weather Services

This manual is designed to help all aviators get the most out of Environment Canada's aviation weather forecasts. It is not intended to be a textbook on aviation meteorology but rather a clear, practical introduction to the aviation weather hazards in British Columbia and the Yukon.

Technology such as satellite imagery, automated weather stations, and computerized information gathering and forecasting systems are all valuable tools used by the aviation forecasters. Despite this technological coverage, in many cases bad weather will remain undetected unless reported by the flying community.

Pilot reports (Pireps) are the backbone to safe flying. All observations are useful, but the most important reports are those which tell of weather conditions that have suddenly worsened or those that greatly differ from the forecast. The observations do not need to be overly detailed and can be as brief as "Rogers Pass closed to VFR traffic due to low cloud." The key is to make the report timely and to be as exact as possible as to the location of the hazard.

Chapter 1

BASICS OF METEOROLOGY

Summer thunderstorm over B.C. Interior

Basics of Meteorology

To properly understand weather, a pilot must understand some of the basic principles that drive the weather machine. There are numerous books on the market that describe these principles in great detail with varying degrees of success. This section is not intended to replace these books, but rather to serve as a review.

Heat Transfer and Water Vapour

The atmosphere is a "heat engine" that runs on one of the fundamental rules of physics, excess heat in one area (the tropics) must flow to colder areas (the poles). There are different methods of heat transfer but by far the most efficient method is through the use of water.

Within our atmosphere, water can exist in three states depending on its energy level. Changes from one state to another are called phase changes and are readily accomplished at ordinary atmospheric pressures and temperatures. The heat taken in or released during a phase change is called latent heat *(see fig. 1)*.

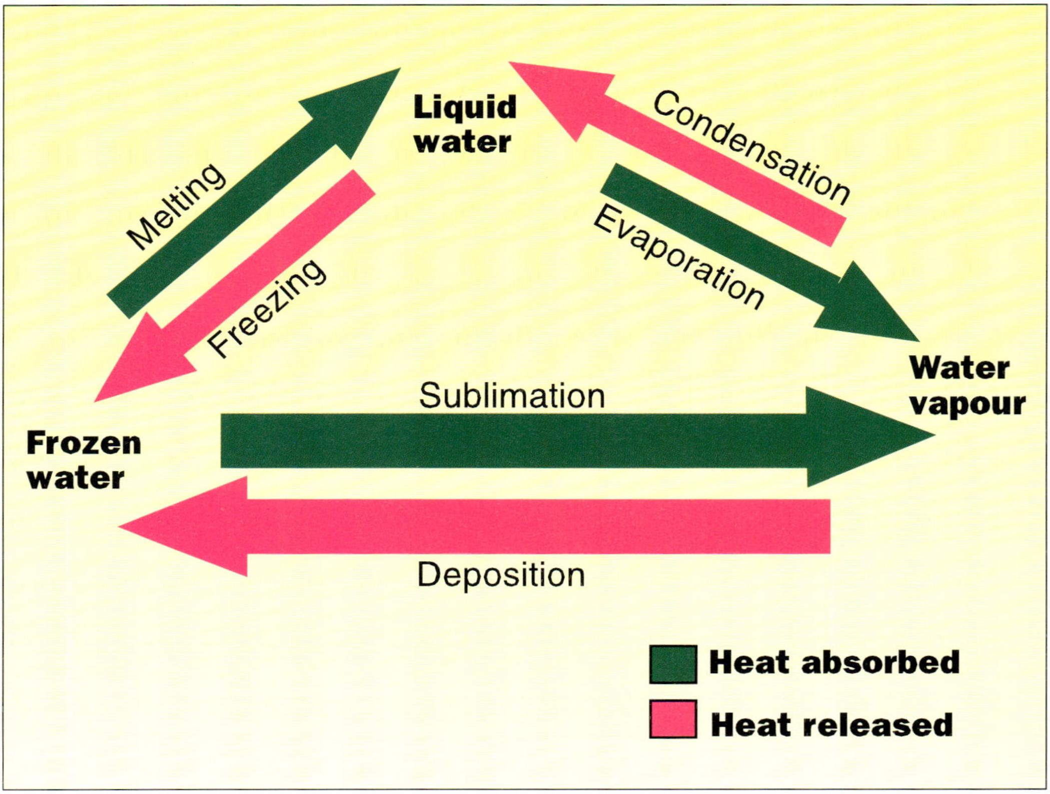

Fig. 1 — Heat transfer and water vapour

How much water the air can hold in the form of vapour is directly related to its temperature. The warmer the air, the more water it can hold. Air that contains the maximum amount of water it can hold, at a given temperature, is said to be saturated and contains maximum latent heat.

The planetary heat engine consists of water being evaporated by the sun into water vapour at the equator (storing heat) and transporting it polewards on the winds where it is condensed back into a solid or liquid state (releasing heat). Most of what we refer to as "weather," such as wind, cloud, fog and precipitation is all related to this conversion activity.

Lifting Processes

The simplest and most common way to get water vapour to convert back to a liquid or solid state is by lifting. When air is lifted, it cools until it becomes saturated. Any additional lift will result in further cooling which reduces the amount of water vapour the air can hold. The excess water vapour is condensed out in the form of cloud droplets or ice crystals which then can go on to form precipitation. There are several methods of lifting an airmass of which the most common are orographic lift (upslope flow), frontal lift, convection and convergence into an area of low pressure *(see figs. 2, 3 and 4)*.

Fig. 2 — Convection as a result of daytime heating

Fig. 3 — Topographical (upslope) lift

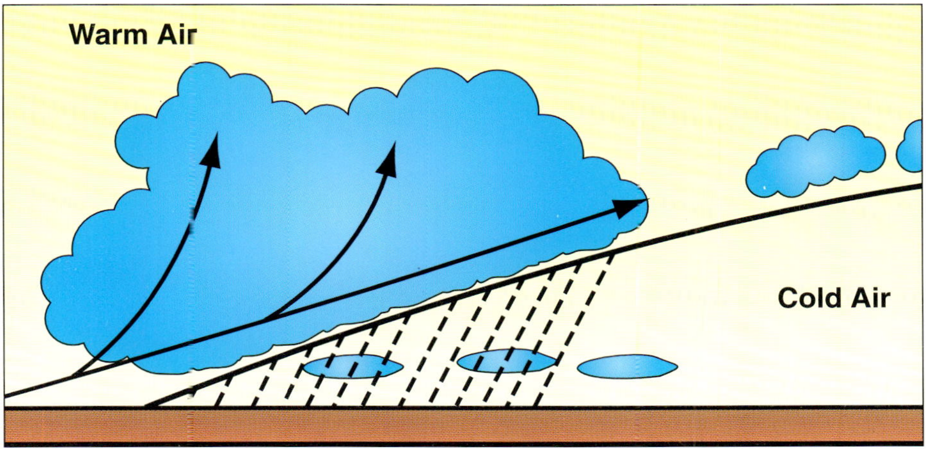

Fig. 4 — Frontal lift

Subsidence

The converse of lifting air is subsiding air in which the air heats as it descends due to compression. The result of this subsidence is that the air is no longer saturated and proceeds to evaporate water. Initially the precipitation ends, followed by the evaporation of cloud droplets and ice crystals. If still unsaturated, the air will then begin to evaporate water from the earth's surface. It is the southwesterly flow over the Coast Mountains subsiding into the interior of British Columbia that creates the arid conditions in the western interior.

An even more classic example, the "Chinook" occurs when air subsides into Alberta during the winter months. In a matter of a few hours temperatures can rise from bitterly cold to shirt sleeve conditions *(see fig. 5)*.

Fig. 5 — Chinook flow

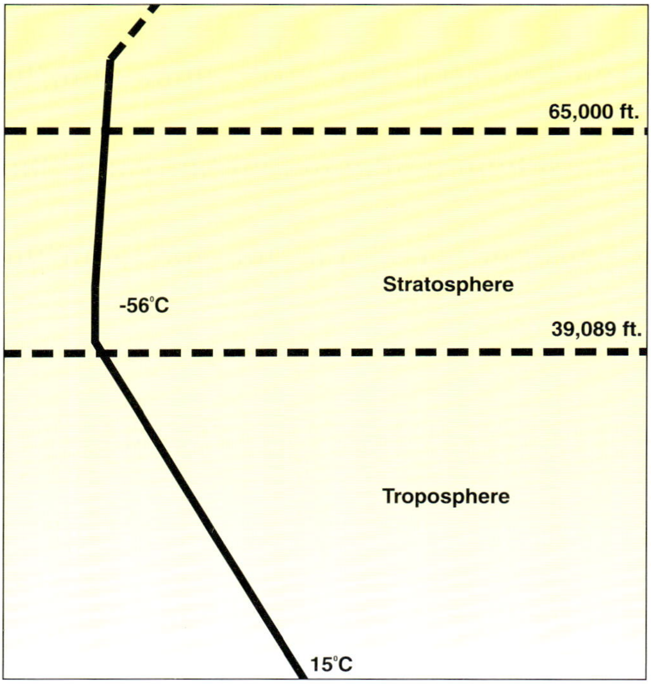

Fig. 6 — ICAO Standard atmosphere

Temperature Structure of the Atmosphere

The temperature lapse rate of the atmosphere refers to the change of temperature with height. In the normal case temperature decreases with height through the troposphere to the tropopause then becomes isothermal in the stratosphere *(see fig. 6)*.

Two other conditions are possible: an inversion in which the temperature increases with height or an isothermal layer in which the temperature remains constant with height *(see fig. 7, opposite)*.

The temperature lapse rate of the atmosphere is a direct measurement of the stability of the atmosphere.

Stability

It would be impossible to examine weather without taking into account the stability of the air. Stability refers to the ability of a parcel of air to resist vertical motion. If a parcel of air is displaced upwards and then released it is said to be unstable if it continues to ascend, stable if it returns to the level from which it originated and neutral if the parcel remains at the level it was released.

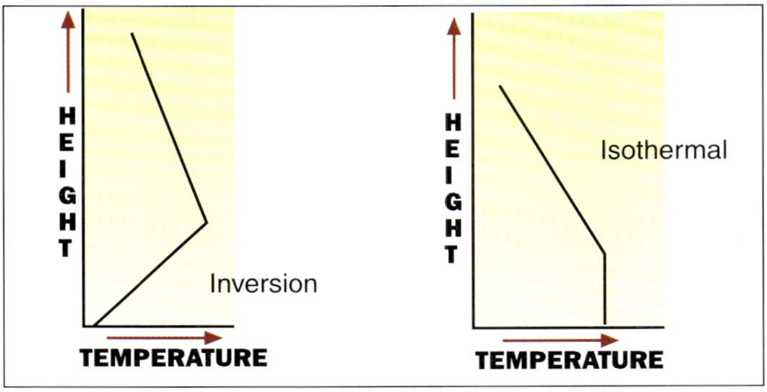

Fig. 7 — Inversion and Isothermal lapse rates

The type of cloud and precipitation produced varies with stability. Unstable air when lifted tends to develop convective clouds and showery precipitation. Stable air tends to produce deep layer cloud and widespread steady precipitation while neutral air tends to produce stable type weather which will change to unstable type weather if the lifting continues *(see fig. 8)*.

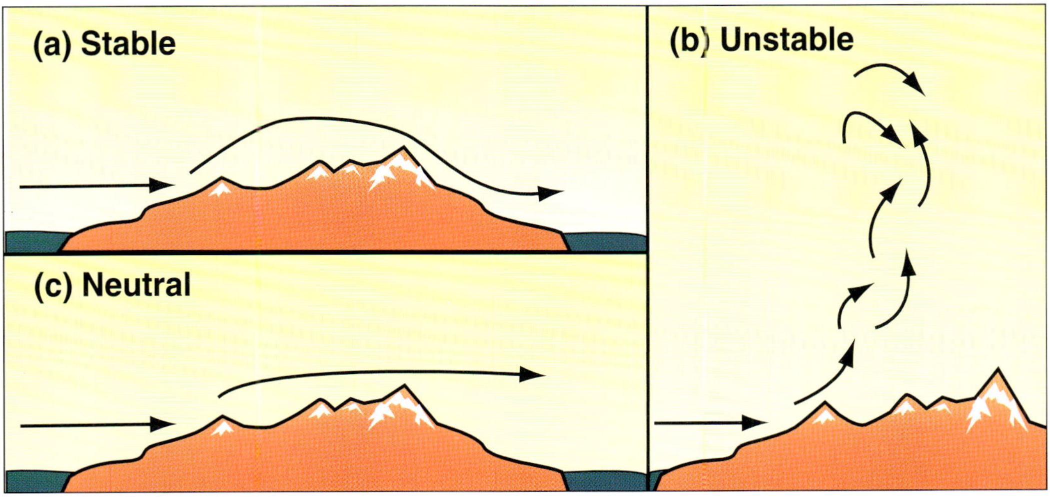

Fig. 8 — Stability in the atmosphere

The stability of an air mass can be changed. One way to destabilize the air is to heat it from below in much the same manner as you would heat water in a kettle. The two easiest ways are to have the sun heat the ground which in turn heats the air in contact with it or by moving cold air over a warmer surface such as open water in winter. The reverse case, cooling the air from below, will stabilize the air. Both processes occur readily. Consider a typical summer day where the air mass is destabilized by the sun, resulting in the development of large convective cloud and accompanying showers or thundershowers during the afternoon and evening. After sunset, the surface cools and the air mass stabilizes slowly, causing the convective activity to die off and the clouds to dissipate.

On any given day there may be several processes acting simultaneously that can either destabilize or stabilize the air mass. To further complicate the issue, these competing effects can occur over areas as large as the entire province to as small as a football field. To determine which one will predominate remains in the realm of the meteorologist and is beyond the scope of this manual.

Wind

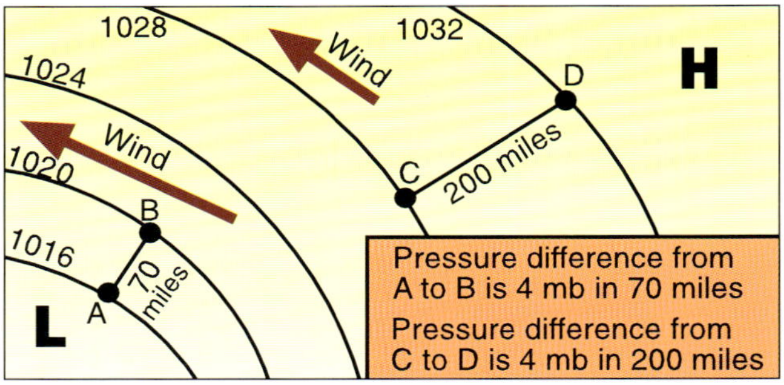

Fig. 9 — The faster pressure changes with horizontal difference, the stronger the wind

Horizontal differences in temperature result in horizontal differences in pressure. It is these horizontal changes in pressure that cause the wind to blow as the atmosphere attempts to equalize pressure by moving air from high pressure to low pressure. The stronger the pressure difference, the stronger the wind and, as a result, the day-to-day wind can range from the gentlest breeze around an airfield to storm force winds over the entire Gulf of Alaska and along the British Columbia coast.

Wind has both speed and direction, so for aviation purposes several conventions have been adopted. Wind direction is always reported as the direction from which the wind is blowing while wind speed is the average steady state value over a certain length of time. Short-term variations in speed are reported as either gusts or squalls depending on how long they last.

Above the surface, the wind tends to be relatively smooth and changes direction and speed only in response to changes in pressure. At the surface the wind is affected by friction and topography. Friction tends to slow the wind over rough surfaces. Topography most commonly induces localized changes in direction and speed.

Air Masses and Fronts

Air Masses

When a section of the troposphere, hundreds of miles across, remains stationary over or moves slowly across an area having fairly uniform temperature and moisture then the air takes on the characteristics of this surface and becomes known as an air mass. The area where air masses are created are called "source regions" and are either ice or snow covered polar regions, cold northern oceans, tropical oceans or large desert areas.

Although the moisture and temperature characteristics of an air mass are uniform, the horizontal weather may vary due to different processes acting on it. It is quite possible for one area to be reporting clear skies while in another area is reporting widespread thunderstorms.

Fronts

When air masses move out of their source regions they come into contact with other air masses. The transition zone between two different air masses is referred to as a frontal zone or front. Across this transition zone temperature, moisture content, and wind can change rapidly over a short distance.

The principal types of fronts are:

Cold Front - The cold air is advancing and overtaking the warm air. The leading edge of the cold air is the cold front.

Warm front - The cold air is retreating and being replaced by warm air. The trailing edge of the cold air is the warm front.

Stationary front - The cold air is neither advancing nor retreating. Frequently referred to quasi-stationary front although in reality there usually is some small-scale localized motion occurring.

Occluded front - More commonly called an "upper front" or "trowal," an occluded front is one that is no longer in contact with the surface.

More will be said about frontal weather later in this manual.

T.Gairns

Thunderstorm gust front at Prince George Airport

Chapter 2

WEATHER HAZARDS TO AVIATION

Ballenas Lightstation, Strait of Georgia *O. Lange*

Introduction

The term "weather" is all encompassing and includes everything from something as simple as the temperature over the tarmac to a severe weather event such as a tornado. Aviation is a specific activity and, although aircraft operate in the realm of "weather," there are only specific elements that offer a potential hazard. In this section we will examine the how and why of some of these hazards.

Icing

One of the oldest assumptions made about clouds is that cloud droplets are in a liquid form at temperatures warmer than 0°C and that they freeze into ice crystals within a few degrees below zero. In reality, 0°C marks the temperature at which water droplets become supercooled and are capable of freezing. Some of the droplets actually do freeze spontaneously just below 0°C while others persist in the liquid state at much lower temperatures.

Aircraft icing occurs when supercooled water droplets strike an airframe whose temperature is colder than 0°C. The effects of the ice on the aircraft can be quite serious and include:

- the disruption of the smooth laminar flow over the wings causing a decrease in lift and an increase in the stall speed.
- an increase in weight and drag thus increasing fuel consumption.
- partial or complete blockage of pitot heads and static ports giving erroneous instrument readings.
- restriction or loss of visibility.

The Freezing Process

When a supercooled water droplet strikes an aircraft surface, it begins to freeze, releasing latent heat. This latent heat warms the remainder of the droplet to near 0°C, allowing the unfrozen part of the droplet to spread back across the surface until freezing is complete. The lower the air temperature and temperature of the aircraft surface, the greater the fraction of the droplet that freezes immediately on impact. Similarly, the smaller the droplet the greater the fraction of the droplet that freezes immediately on impact. Finally, the more frequent the droplets strike the aircraft surface, the greater the amount of water to flow back over the aircraft surface. In general, the maximum potential for icing occurs with large droplets at temperatures just below 0°C.

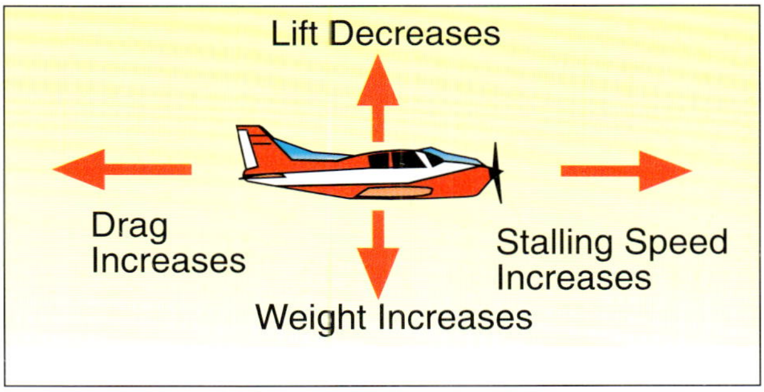

Fig. 10 — Effects of icing on aircraft

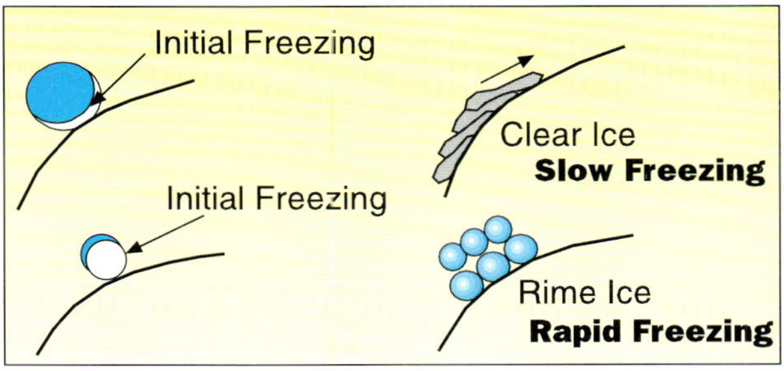

Fig. 11- Freezing of super-cooled droplets on impact

Types of Aircraft Ice

Rime Ice

Rime ice is a product of small droplets where each droplet has a chance to freeze completely before another droplet hits the same place. The ice that is formed is opaque and brittle because of the air trapped between the droplets. Rime ice will usually only form on the leading edges of airfoils, builds forward into the airstream and has low adhesive properties.

Clear Ice

In the situation where each large droplet does not freeze completely before additional droplets hit the same place, then the supercooled water from each drop merges and spreads backwards across the aircraft surface before freezing completely to form an ice with high adhesive properties. Clear ice tends to range from transparent to a very tough opaque layer and will build back across the aircraft surface as well as forward into the airstream.

Mixed Ice

When the temperature and the range of droplet size varies widely, the ice that forms is a mixture of rime ice and clear ice. This type of ice tends to have more adhesive properties than rime ice, is opaque in appearance, rough, and generally builds forward into the airstream faster than it spreads back over the aircraft surface *(see fig. 12)*.

Meteorological Factors Affecting Icing

(a) Liquid Water Content of the Cloud

The liquid water content of a cloud is dependent on the size and number of droplets in a given volume of air. The greater the liquid water content, the more serious the icing. Clouds with strong vertical updrafts generally have a higher liquid water content as the updrafts prevent even the large drops from precipitating.

The strongest updrafts are to be found in convective clouds, clouds formed by abrupt orographic lift and in lee wave clouds. Stratiform clouds tend to have weak updrafts and are generally composed of small droplets.

(b) Temperature Structure in the Cloud

Warm air can hold more water vapour than cold air. Thus, clouds that form in warm

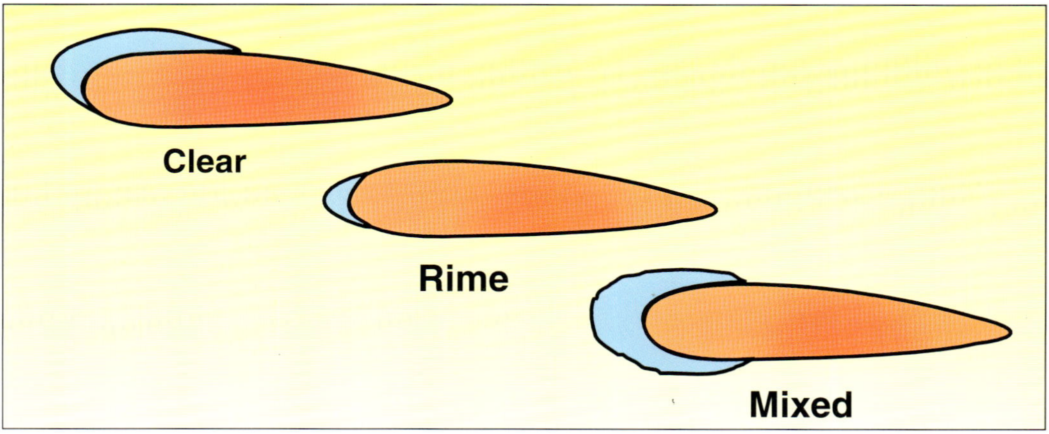

Fig. 12 — Accumulation patterns of different icing type

air masses will have a higher liquid water content than those that form in cold air.

The temperature structure in a cloud has a significant effect on the size and number of droplets. The larger supercooled droplets begin to freeze spontaneously around -10°C with the rate of freezing of all size of droplets increasing rapidly at temperatures just below minus 15°C. By -40°C, virtually all the droplets will have frozen. The exceptions are clouds with very strong vertical updrafts, such as towering cumulus or cumulonimbus, where liquid water droplets can be carried to great heights before freezing.

All these factors can cause the icing intensity to change rapidly with time so that it is possible for aircraft following each other by only a few minutes to encounter different icing conditions. However, some generally accepted rules have been developed *(see fig. 13)*.

(1) Within large cumulus and cumulonimbus clouds:

- at temperatures between 0°C and -25°C, severe clear icing likely.
- at temperatures between -25°C and -40°C, light rime icing likely; small possibility of moderate to severe rime or mixed icing in newly developed clouds.
- at temperatures below -40°C, little chance of icing.

(2) Within layer cloud:

- the icing layer is generally confined to the 0°C to -15°C temperature range.
- icing is usually less severe than convective cloud because of the weaker updrafts and smaller droplets.
- icing layers tend to be shallow in depth but great in horizontal extent.

(3) Situations in which icing may be greater than expected:

- air moving across large unfrozen lakes in the winter will increase its moisture content and destabilize rapidly due to heating from below. The cloud that forms, while resembling a layer cloud, will actually be a convective cloud capped by an inversion with relatively strong updrafts and a large concentration of super-cooled drops.
- thick layer cloud formed by rapid mass ascent such as in an intensifying low or along

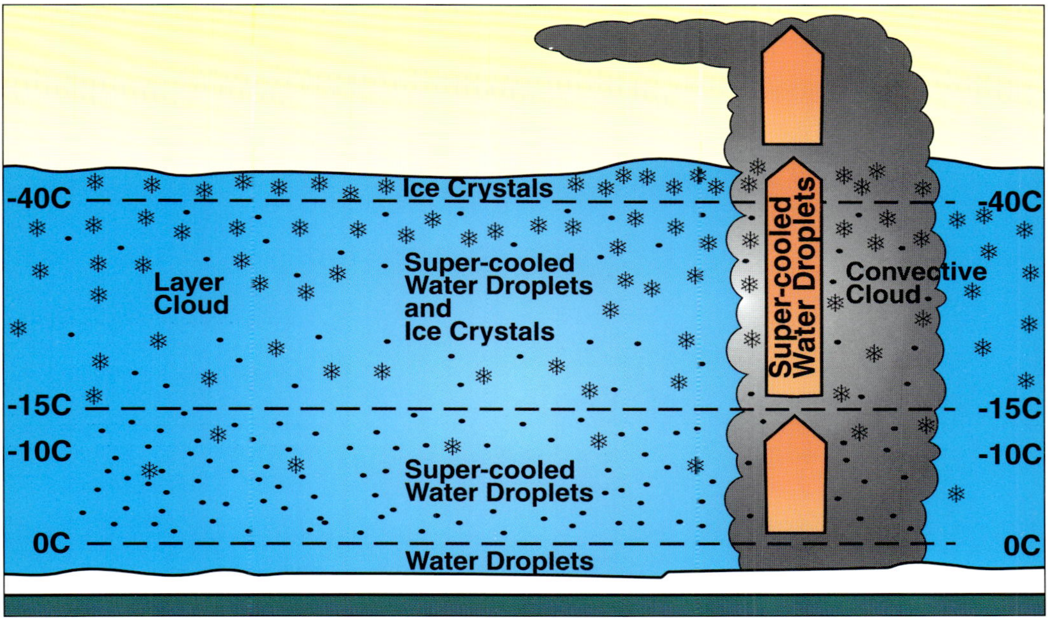

Fig. 13 — Distribution of water droplet-ice crystals in cloud

mountain slopes will also have enhanced concentrations of supercooled drops. There is also a strong possibility that such lift will destabilize the airmass resulting in embedded convective clouds with their enhanced icing potential.

- lenticular clouds can have very strong vertical currents associated with them. Icing can be severe and, because of the droplet size, tend toward clear icing.

Aerodynamic Factors Affecting Icing

There are various aerodynamic factors that affect the collection efficiency of an aircraft surface. Collection efficiency can be defined as the fraction of liquid water that actually strike the aircraft relative to the number of droplets encountered along the flight path.

Collection efficiency is dependent on three factors:

$$\text{collection efficiency} = \frac{\text{Mass of supercooled water droplets that stick to the airfoil}}{\text{Mass of supercooled droplets encountered by the airfoil}}$$

(a) The radius of curvature of the aircraft component. Components with a big radius of curvature disrupt the airflow (like a bow wave) causing the smaller supercooled droplets to be carried around the airfoil by the airstream. For this reason, large thick components (thick wings, canopies) collect ice less efficiently than thin components (thin wings, struts, antenna).

(b) Speed. The faster the aircraft the less chance the droplets have to be diverted around the aircraft by the airstream.

(c) Droplet size. The larger the droplet the more difficult it is for the airstream to displace it *(see fig. 14)*.

Other Forms of Icing

(a) Freezing Rain

Freezing rain occurs when liquid water drops fall into a layer of air whose temperature is colder than 0°C and supercool before hitting some object. The most common scenario leading to freezing rain in British Columbia and the Yukon is "warm overrunning." In this case, cold air fills the valleys. While aloft, the southwesterly winds are spreading warm air, cloud and precipitation into the province. In such a scenario, rain that falls into the cold air supercools, giving freezing rain that can last for hours. Such a condition is relatively common in the inlets of the British Columbia coast.

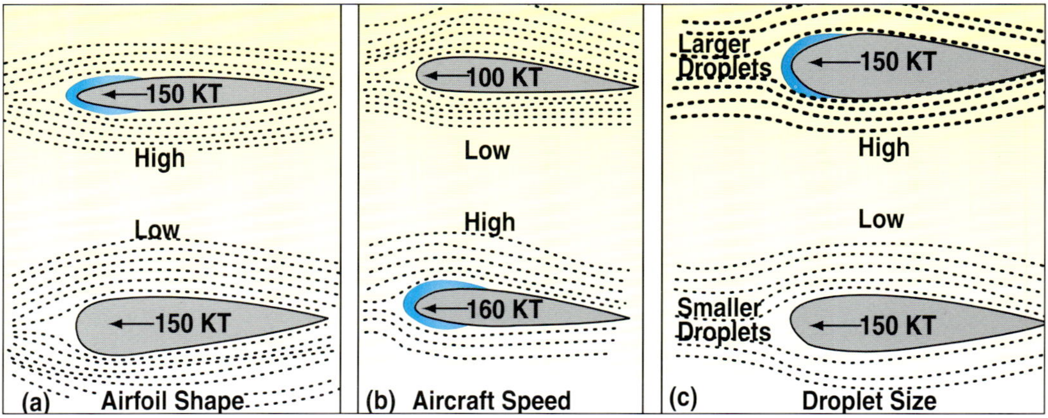

Fig. 14 — Variations in collection efficiency

(b) Snow

Dry snow will not adhere to an aircraft surface and will not normally cause icing problems. Wet snow however can freeze hard to an aircraft surface that is at subzero temperatures and be extremely difficult to remove.

A very dangerous situation can arise when an aircraft attempts to take off with wet snow on the flight surfaces. Once the aircraft is set in motion, evaporational cooling will cause the wet snow to freeze hard causing a drastic reduction in lift as well as increasing the weight and drag. Wet snow can also freeze to the windscreens making visibility difficult to impossible.

(c) Freezing Spray

Freezing spray develops over the coastal waters of British Columbia when there is a strong outflow of arctic air through the inlets. While the water itself is above freezing, any water that is picked up by the wind or splashes onto an object will quickly freeze, causing a rapid increase in weight and shifting the centre of gravity. Such conditions have caused one weather buoy on Nanakawa Shoals in Douglas Channel to completely overturn. Another weather buoy anchored in Howe Sound was completely encrusted in ice. Similar problems could be encountered by seaplanes or float-equipped planes.

Visibility

Visibility is the meteorological component which most impacts flight operations. Mountainous valleys all tend to look the same at low levels making good route navigation essential. This can only be done in times of clear visibility.

Reduced visibility near Kicking Horse Pass in August *K. Johnson*

Types of Visibility

There are several terms used to describe the different types of visibility used by the aviation community.

(a) Horizontal visibility — the furthest visibility obtained horizontally in a specific direction by referencing objects or lights at known distances.
(b) Prevailing visibility — the ground level visibility that which is common to one-half or more of the horizon circle.
(c) Vertical visibility — the maximum visibility obtained by looking vertically upwards into a surface-based obstruction such as fog or snow.
(d) Air-to-ground visibility — Visibility observed by looking downwards from the cockpit of the aircraft.

Causes of Reduced Visibility

(a) Lithometers

Lithometers are dry particles suspended in the atmosphere and include haze, smoke, sand and dust. Of these, smoke causes the most problems in British Columbia and the Yukon. The common sources of smoke are slash burning fires and forest fires. Smoke from distant sources will resemble haze, but near a fire, smoke can reduce visibility to below VFR conditions.

(b) Precipitation

Rain can reduce visibility, however, the restriction is seldom less than one mile other than in the heaviest showers beneath cumulonimbus cells. Drizzle, because of the greater number of drops in each volume of air, is usually more effective at reducing the visibility especially when accompanied by fog.

Snow affects visibility more than rain or drizzle and can easily reduce it to less than one mile. Blowing snow is a product of strong winds picking up the snow particles into the air. Fresh fallen snow is easily disturbed and can be lifted a few hundred feet. Under extreme conditions, the cockpit visibility will be excellent until the aircraft "rounds out" at which time the horizontal visibility will be reduced abruptly.

(c) Fog

Fog is the most common and persistent visibility obstruction encountered by the aviation community.

A cloud based on the ground, fog can consist of water droplets, supercooled water droplets, ice crystals or a mix of the supercooled droplets and ice crystals. The names given to the types of fog (radiation, upslope, advection, etc.) are all based on how they are formed. All types are capable of forming rapidly, restricting both the ceiling and visibility to near zero.

Wind, Shear and Turbulence

The "why" of the winds are quite well understood. It is the daily variations of the winds, where they blow and how strong, that remains a constant problem for meteorologists to unravel. The problem becomes even more difficult when local effects such as wind flow through coastal inlets or in mountain valleys is added to the dilemma. The result of these effects can give one airport persistent light winds while another has nightly episodes of strong gusty winds.

Stability and the Diurnal Variation in Wind

In a stable weather pattern, daytime winds are generally stronger and gustier than the nighttime winds. During the day, the heating from the sun sets up convective mixing which carries the stronger winds aloft down to the surface and mixes it with the slower surface winds. This causes the surface winds to increase in speed and to become gusty especially during the late afternoon when maximum heating occurs.

After sunset, the surface of the earth cools which in turn cools the air near the surface resulting in the development of a temperature inversion. This inversion deepens as cooling continues, bringing an end to convective mixing and causes the surface winds to slacken.

Wind Shear

Wind shear is nothing more than a change in wind direction and/or wind speed over the distance between two points. If the points are in a vertical direction then it is called vertical shear, if they are in a horizontal direction than it is called horizontal shear.

In the aviation world, the major concern is how abruptly the change occurs. If the change is gradual, a change in direction or speed will result in nothing more than a minor change in the ground speed. If the change is abrupt, there will be a rapid change of airspeed or track. Depending on the aircraft type, it may take a significant time to correct the situation placing the aircraft in peril particularly during takeoff and landing.

Significant shearing can occur when the wind being channeled along a valley varies significantly from the free flowing wind above the valley. Changes in direction of 90° and speed changes of 25 knots are reasonably common in mountainous terrain. Kamloops airport is a good example of this condition *(see fig. 15)*.

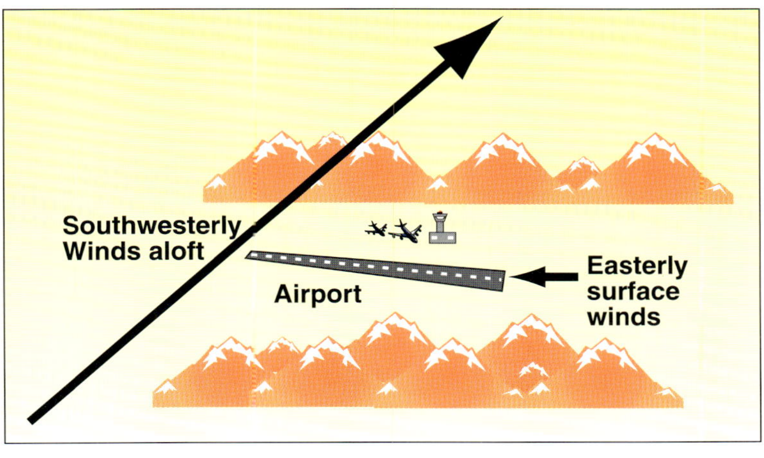

Fig. 15 — Wind shear over a valley

Shears are also induced by updrafts and downdrafts. An abrupt downdraft will cause a brief decrease in the wing's attack angle resulting in a loss of lift. An updraft will increase the wing's attack angle increasing the lift however there is a risk that it could be increased beyond the stall angle.

Shears can also be encountered along fronts. Frontal zones are generally thick enough that the change is gradual, however cold frontal zones as thin as 200 feet have been measured. Strong shears are unusual with fronts but pilots doing a take-off or a let-down through a frontal surface near the ground should be wary.

Mechanical turbulence is a form of shear induced when a rough surface disrupts the smooth wind flow. The amount of shearing and the depth of the shearing layer depends on the wind speed, the roughness of the obstruction and the stability of the air.

Wind Shear and Turbulence

Turbulence is the direct result of wind shear tending shear zones to be turbulent. The stronger the shear the stronger the turbulence. However, not all shear zones are turbulent, so the absence of turbulence does not infer that there is no shear.

Topographical Effects on Wind

(a) Lee effects

When the winds blow against a steep bluff or over rugged terrain, gusty turbulent winds result. Eddies often form downwind of the cliff face which create stationary zones of stronger and lighter winds. These zones of strong winds are fairly predictable and usually persist as long as the wind direction and stability of the airstream do not change. The lighter winds, which occur in areas called wind shadows, can vary in speed and direction, particularly downwind of higher cliffs. An example is the bluffs north of Prince George airport.

Beneath the cliffs the wind is usually gusty and the wind direction is often completely opposite to the wind blowing over the top of the cliff. Smaller, reverse eddies may also be encountered close to the cliffs *(see fig. 16)*.

(b) Friction effects

The winds that blow well above the surface of the earth are not strongly influenced by

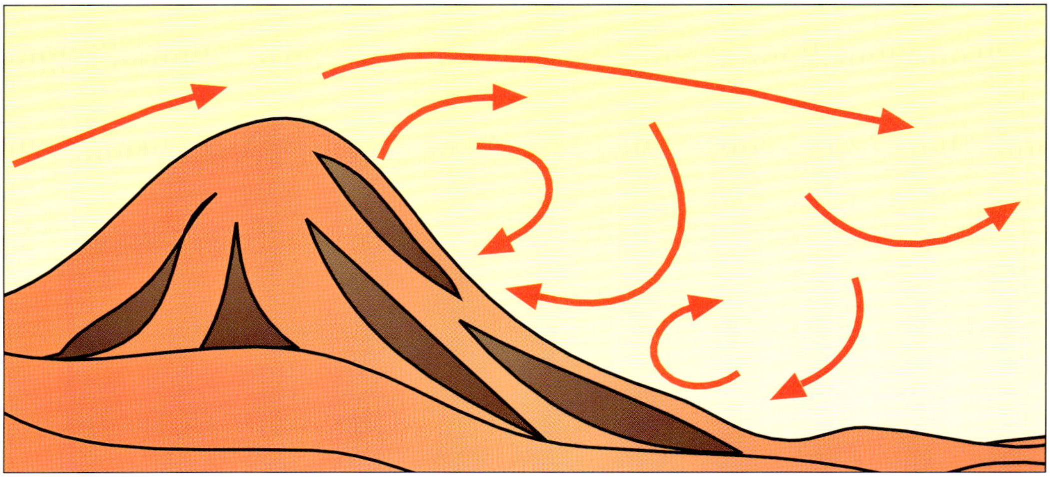

Fig. 16 — Lee effects

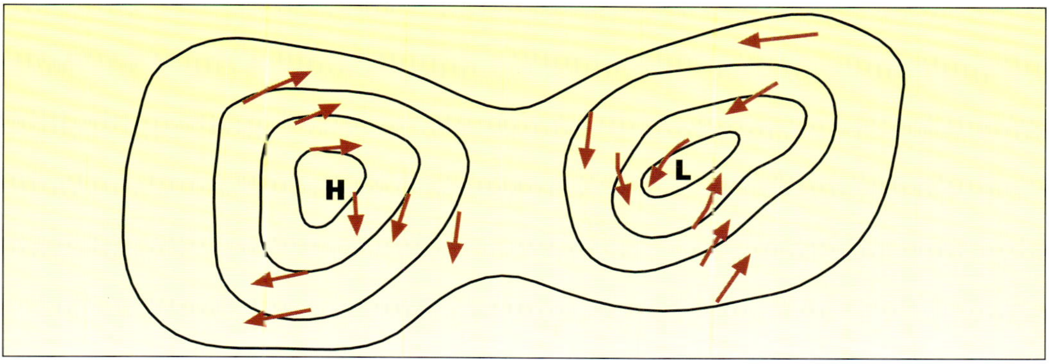

Fig. 17 — Friction effect: Wind direction (surface winds)

the presence of the earth itself. Closer to the earth, frictional effects decrease the speed of the air movement and back the wind (turns the wind direction counterclockwise) towards the lower pressure. For example, a southerly wind becomes more southeasterly when blowing over rougher ground. Since both British Columbia and the Yukon are very rugged, there is a significant reduction in the wind speed over the land when compared to the wind produced by the same pressure gradient over the relatively smooth sea *(see fig. 17)*.

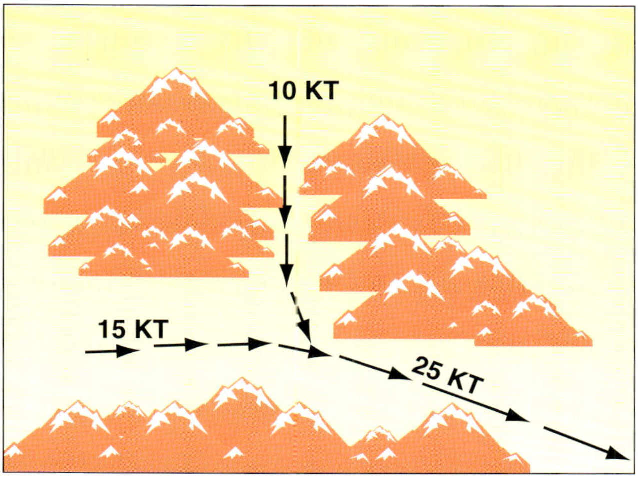

Fig. 18 — Converging winds

(c) Converging winds

When two or more airstreams flow together or converge the effect is to create a stronger stream. Similar effects can be noted where two or more valleys come together, such as near Nelson *(see fig. 18)*. Another example of this occurs along the coast when the different angles of the surface winds over land and water cause the airstreams to converge. This convergence creates a band of wind which is about 25 percent stronger within a few miles of the shore.

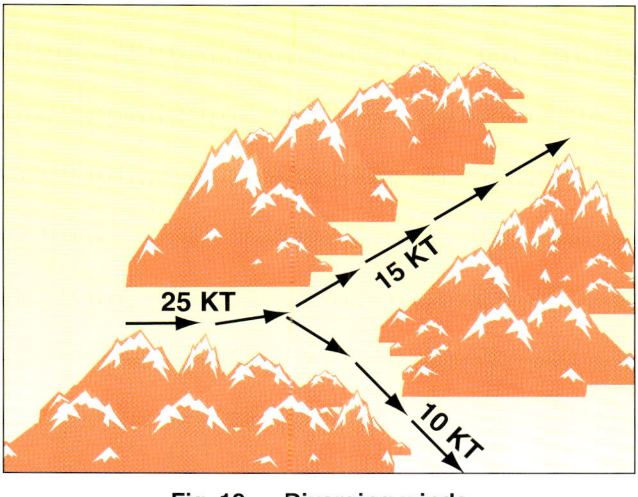

Fig. 19 — Diverging winds

(d) Diverging winds

A divergence of the airstream occurs when a single airstream splits into two or more streams. Each will have a lower speed than the parent airstream. This is common near Lillooet *(see fig. 19)*.

(e) Corner winds

When the prevailing wind encounters a headland, there is a tendency for the wind to curl around the feature. This change in direction, if done abruptly, can result in extreme turbulence. An example of this is near Telegraph Point, east of Prince Rupert *(see fig. 20)*.

Fig. 20 — Corner Winds

(f) Funneled winds

When winds are forced to flow through a narrow opening or gap, such as an inlet or narrow section of a pass, the wind speed will increase and may even double in strength. This effect is similar to pinching a water hose and is called funneling. The Penticton airport, in a narrow portion of the Okanagan Valley, is a prime example *(see fig. 21)*.

(g) Channeled winds

The topography can also change the direction of the winds by forcing the flow along the direction of a pass or through a strait. This is referred to as channeling. For example, winds at Kamloops are almost always easterly or westerly, along the South Thompson River Valley. Channeled winds are also common through Juan de Fuca Strait *(see fig. 22)*.

Fig. 21 — Funnelled winds

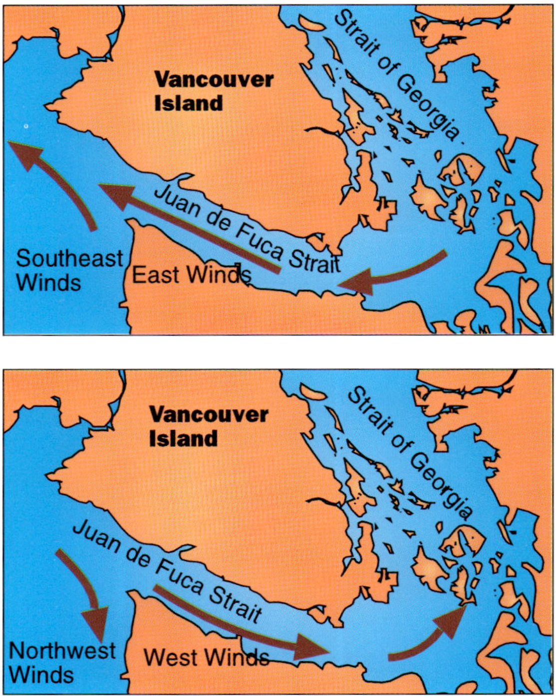

Fig. 22 — Channeled Wind

(h) Land and Sea breezes

Land and Sea Breezes are only observed under light wind conditions and depend on temperature differences between adjoining regions.

The sea breeze occurs when the air over the land is heated more rapidly than the air over the adjacent water surface. As a result, the warmer air rises and the relatively cool air from the water flows onshore to replace it. By late afternoon, the time of maximum heating, the seabreeze circulation may reach speeds of 10 to 15 knots and extend 15 miles inland.

During the evening the sea breeze subsides. At night, as the land cools, a land breeze develops in the opposite direction and flows from the land out over the water. It is generally not as strong as the sea breeze, but at times it can be quite gusty.

Both the land and sea breeze can be influenced by channeling and funneling resulting in almost frontal-like conditions with sudden wind shifts and gusty winds that may reach up to 40 knots, particularly in coastal inlets such as Howe Sound *(see fig. 23)*.

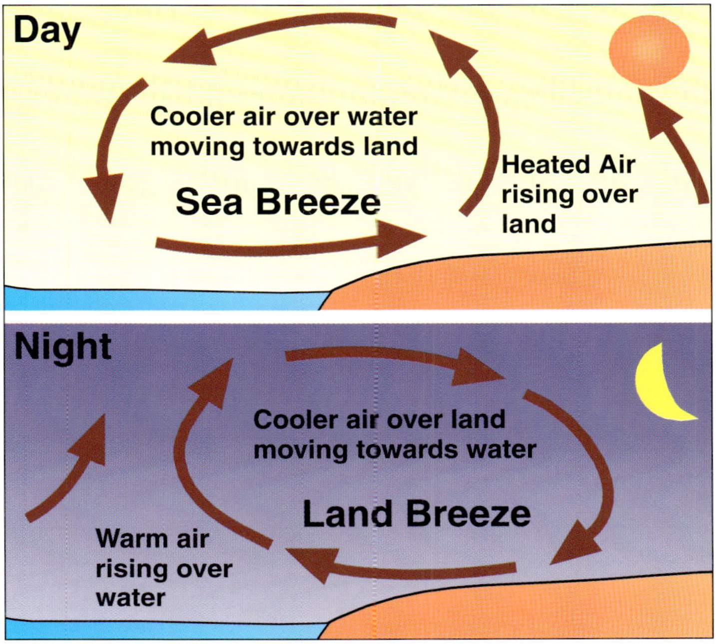

Fig. 23 — Land and sea breezes

(i) Katabatic and Anabatic winds

During the day the sides of the valleys become warmer than the valley bottoms since they are better exposed to the sun. As a result, the winds blow up the slopes. These daytime, upslope winds are called anabatic winds. Gently sloped valley sides, especially those facing south, are more efficiently heated than those of a steep, narrow inlet. As a result, valley breezes will be stronger in the wider valleys. An anabatic wind, if extended to sufficient height, will produce cloud. In addition, such a wind offers additional lift to aircraft and gliders.

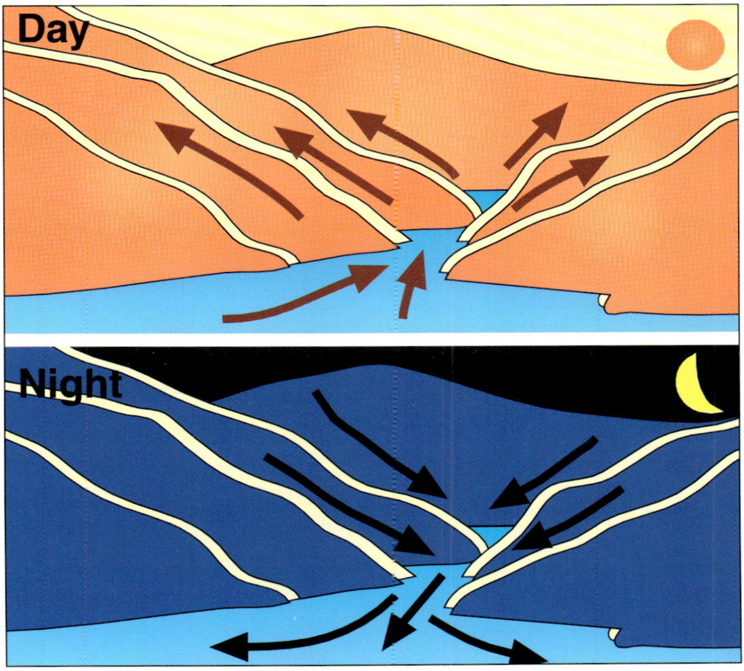

Fig. 24 — Anabatic and katabatic winds

At night, the air cools over the mountain slopes and sinks to the valley floor. If the valley floor is sloping, the winds will move along the valley towards lower ground. The cool night winds are called drainage winds or katabatic winds and are often quite gusty and usually stronger than the daytime anabatic winds. Some mountain valley airports have wind socks situated at various locations along their runways to show the changeable conditions due to the katabatic flow. Strong katabatic winds are observed frequently in many locales, including the Bull River Valley in southeastern British Columbia *(see fig. 24, previous page)*.

(j) Glacier Winds

Under extreme cooling conditions, such as an underlying ice cover, the katabatic winds can develop to hazardous proportions. As the ice is providing the cooling, a shallow wind of 80 knots or more can form and will persist during the day and night. In some locations the katabatic flow "pulsates" with the cold air building up to some critical value before being released to rush downslope. Such winds tend to be extremely violent and values in excess of 100 knots have been observed along the Slims River in the Yukon.

• • •

It is important to recognize that combinations of these effects can operate at any given time. Katabatic winds are easily funneled or channeled resulting in winds of unexpected directions and strengths in narrow passes. Around glaciers in the summer, wind fields can be chaotic as katabatic winds from the top of the glacier struggle for dominance with localized convection or anabatic winds induced by heated rock slopes below the ice. Many sightseeing pilots prefer to avoid glaciated areas during the afternoon hours.

D. Rothwell

Location where glacier winds might be encountered

Lee Waves

When air flows across a mountain or hill, it is disturbed the same way that water flowing over a rock is disturbed. The air initially is displaced upwards across the mountain, dips sharply on the lee side, then rises and falls in a series of waves downstream. These waves are called "mountain waves" or "lee waves" and are most notable for their turbulence. Since mountain ranges can have a greater effect than individual mountains, British Columbia and the Yukon have ideal topography for lee waves.

The Formation of Lee Waves

The development of lee waves requires that several conditions be met:

(a) the wind direction must be within 30 degrees of perpendicular to the mountain or hill. In British Columbia, this would be a wind from the southwest to west. The greater the height of the mountain and the sharper the drop off to the lee side, the more extensive the induced oscillations *(see fig. 25)*.

(b) wind speeds must exceed 15 knots for small hills and 30 knots for mountain ridges. A jet stream with its associated strong winds below the jet axis is an ideal situation.

(c) the wind should be constant direction and increase in speed with height throughout the troposphere.

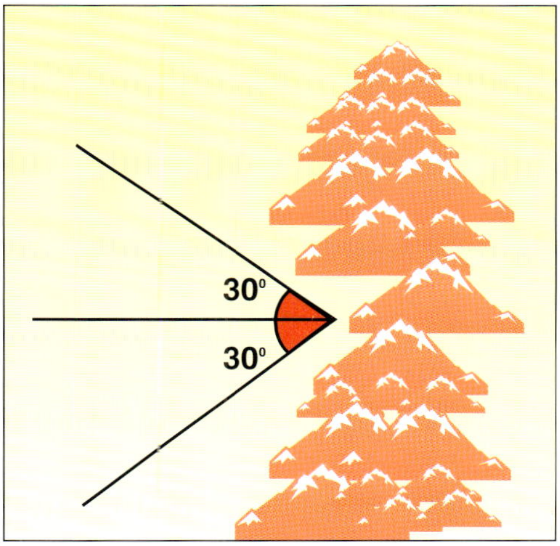

Fig. 25 — Direction cone to produce lee waves in cross-mountain flow

(d) the air should be stable near the mountain peaks but less stable below. The unstable layer encourages the air to ascend and the stable layer encourages the development of a downstream wave pattern.

While all these conditions can be met at any time in the year, winter wind speeds are generally stronger resulting in stronger and more dangerous lee waves.

Characteristics of Lee Waves

Once a lee wave pattern has been established, it follows several basic rules:

- the stronger the wind, the longer the wavelength. The typical wavelength is about 6 miles but they can vary from as short as 3 miles to as long as 15 miles.
- the position of the waves crests will remain nearly stationary with the wind blowing through them as long as the mean wind speed remains nearly constant.
- the wave amplitude can exceed 3,000 feet *(see fig. 26, next page)*.
- the induced vertical currents within the wave can reach values of 4,500 feet per minute.

- the wind speed is stronger through the wave crest and slower through the wave trough *(see fig. 27)*.
- the wave closest to the obstruction is usually the strongest and they get progressively weaker downstream.
- a large eddy called a "rotor" may form below each wave crest *(see fig. 28)*.
- mountain ranges downstream may amplify or nullify induced wave patterns.

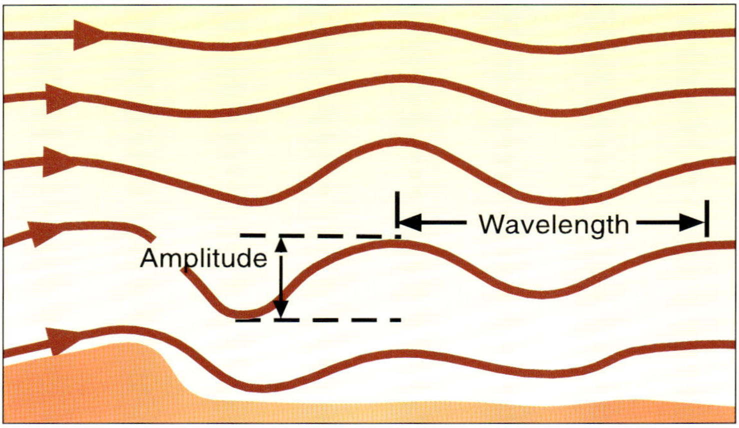

Fig. 26 — Amplitude and wavelength in lee waves

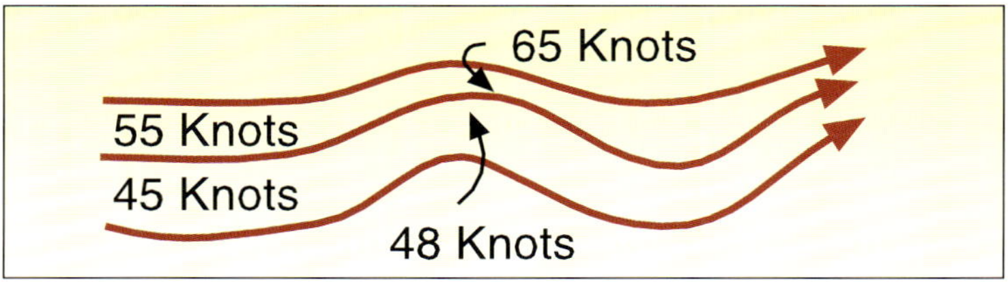

Fig. 27 — Stronger wind in wave crest in lee waves

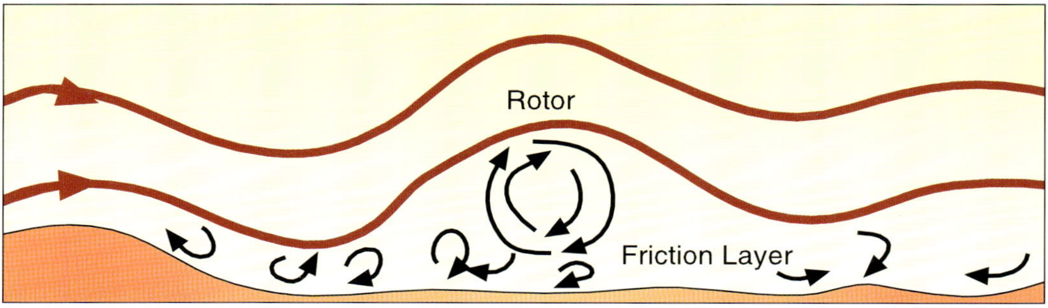

Fig. 28 — The rotor may form beneath wave crests

Clouds Associated with Lee Waves

Lee waves involve lift and, if sufficient moisture is available, characteristic clouds will form. However, the signature clouds may be absent due to the air being too dry or the cloud could be embedded within other clouds and not visible. *It is essential to realize that the absence of lee wave clouds does not mean that there are no lee waves present.*

(a) Cap cloud

A cloud often forms over the peak of the mountain range and remains stationary. Frequently, it may have an almost "waterfall" appearance on the leeward side of the mountain. This effect is caused by subsidence and often signifies a strong downdraft just to the lee of the mountain top.

(b) Lenticular clouds

A lens shaped cloud may be found at the crest of each wave. These clouds may be separated vertically with several thousand feet between each cloud or may form so close together they resemble a "stack of plates." When air flows through the crest it is often laminar making the cloud smooth in appearance.

On occasion, the shear is too much and turbulence results causing the lenticular cloud to appear to be ragged and wind torn.

(c) Rotor cloud

A rotor cloud may form in association with the rotor. It will appear as a long line of stratocumulus, a few miles downwind and parallel to the ridge. Its base will be normally below the peak of the ridge, but its top can extend above. The turbulence associated with a rotor cloud is severe within and near the rotor cloud *(see fig. 29)*.

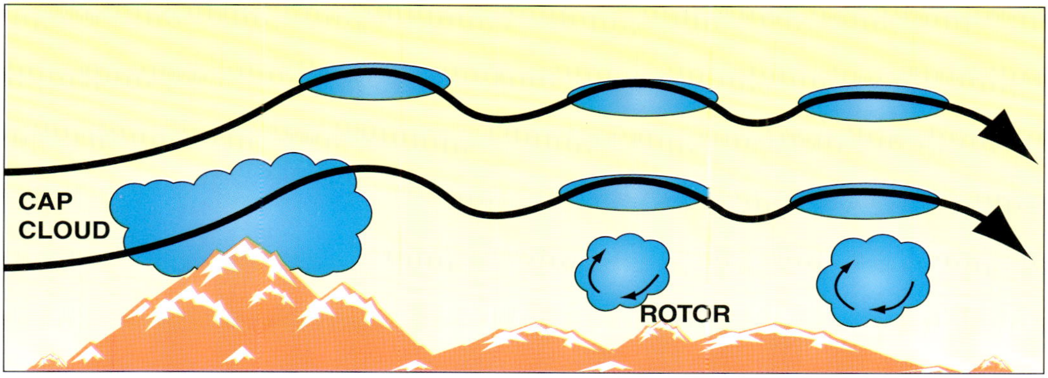

Fig. 29 — Characteristic clouds formed by lee waves.

Active cumulonimbus (CB) clouds *K. Johnson*

Thunderstorms

No weather encountered by a pilot can be as unpredictable or as violent as a thunderstorm. There are few severe weather hazards that a thunderstorm does not pack, and with thunderstorms being common, it is important that a pilot understand a little about their nature and how to deal with them.

The Life Cycle of a Thunderstorm

The thunderstorm, which may cover an area ranging from 5 miles in diameter to as much as 50 miles in the extreme case, usually consists of two or more cells in different stages of their life cycle. The stages of life of individual cells are *(see fig. 30)*:

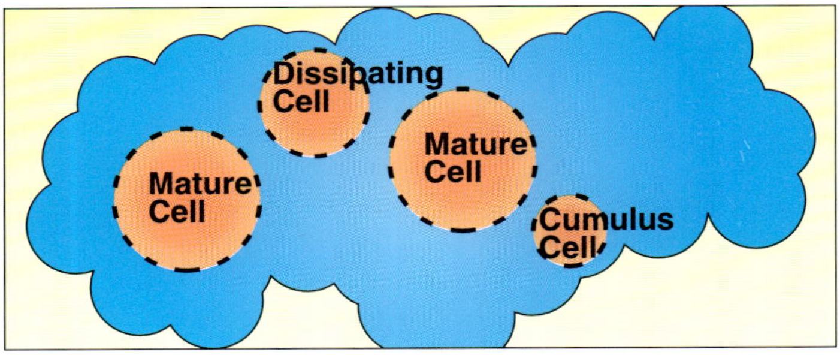

Fig. 30 — Thunderstorm cells in different stages of development

(a) Cumulus Stage

The cumulus stage is marked by updrafts only within the cloud. Consisting of cumulus (CU) and towering cumulus clouds (TCU), the updraft builds the cloud vertically and carries super-cooled water droplets well above the freezing level. Near the end of this stage, the cloud may well have a base more than 5 miles across and a vertical extent in excess of 20,000 feet. The average life of this stage is about 20 minutes *(see fig. 31)*.

(b) Mature Stage

The transition to this stage is marked by the appearance of precipitation beneath the base of the cell and the development of the downdraft. The downdraft is caused by water drops which have become too heavy for the updraft to support and now begin fall. At the same time, the drops begin to evaporate as they move into drier air beneath the base of the cloud. This evaporation causes the air to cool and become more dense resulting in a downwash of accelerating cold air. Downdraft speeds can reach values of 2,500 feet per minute.

The downdraft, when it hits the ground, spreads out in all directions but fastest in the direction the storm is moving. The leading edge of this cold air is called the "gust front" and can extend ten to fifteen miles in front of the storm or even farther when channeled down mountain valleys. This horizontal flow of gusty surface winds is characterized by a rapid drop in temperature and a sharp rise in pressure.

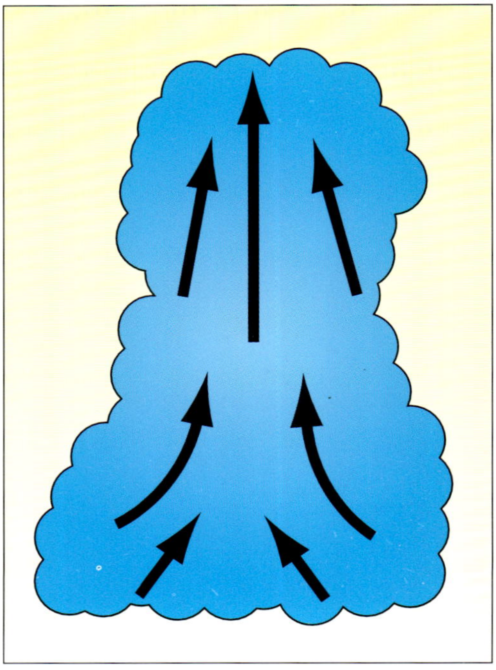

Fig. 31 — Cumulus stage

At the same time, the updrafts continue to strengthen until they reach maximum speed, possibly exceeding 6,000 feet per minute. The cloud reaches the tropopause which blocks the updraft, forcing the stream of air to spread out horizontally. Strong upper winds at the tropopause level assist in the spreading out of this flow in the downwind direction producing the traditional anvil-shaped top. This is classically what is referred to as a cumulonimbus cloud (CB).

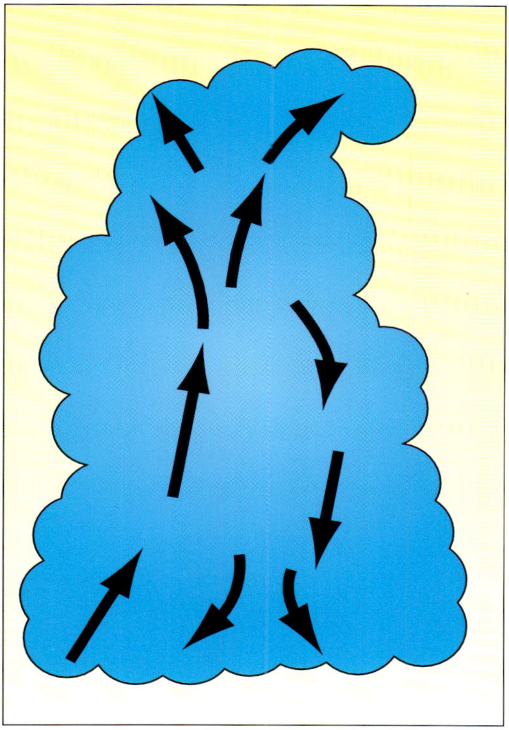

Fig. 32 — Mature stage

In British Columbia, the thunderstorm may have a base measuring from near 5 miles to more than 15 miles in diameter and a top ranging from 25,000 to 45,000 feet (over the Great Plains, tops in excess of 50,000 feet are common). The mature stage is the most violent stage in the life a thunderstorm and usually last for 20 to 30 minutes *(see fig. 32)*.

Near the end of the mature stage, the downdraft has increased in size so that the updraft is almost completely "choked off," stopping the development of the cell. However, at times, the upper winds increase strongly with height causing the cell to tilt. In such a case the falling precipitation falls through only a portion of the cell allowing the updraft to persist and reach values of 10,000 feet per minute. Such cells are referred to as "steady state storms" that can last for several hours and produce the most severe weather including tornadoes *(see fig. 33)*.

(c) Dissipating Stage

The dissipating stage of a cell is marked by the presence of downdrafts only. With no additional flow of moisture into the cloud from an updraft, the rain gradually tapers off and the downdrafts weaken. The cell may dissipate completely in 15 to 30 minutes leaving clear skies or patchy cloud layers *(see fig. 34, next page)*.

Specific Hazards

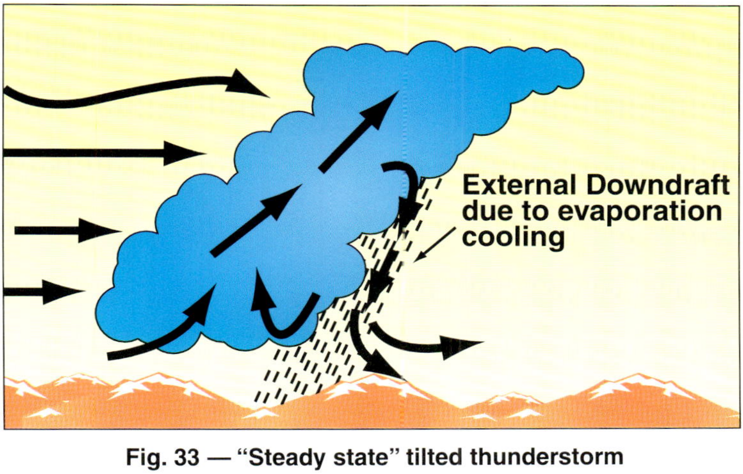

Fig. 33 — "Steady state" tilted thunderstorm

The environment in and around a thunderstorm can be the most hazardous encountered by an aircraft. In addition to the usual risks such as severe turbulence, severe clear icing, large hail, heavy precipitation, low visibility and electrical discharges within and near the cell, there are other hazards that occur in the surrounding environment.

(a) The Gust Front

The gust front runs out ahead of the storm by as much as 15 miles and may occur under relatively clear skies. Aircraft taking off, landing, or operating at low levels can find themselves in a rapidly shifting wind field in which the wind direction can change as much 180° in a matter of a few seconds and wind speed accelerate to values of 25 to 40 knots or more with stronger gusts. Extremely strong gust fronts can do considerable damage on the ground and are sometimes referred to as "plow winds." All of this will likely be accompanied by considerable mechanical turbulence and induced shear on the frontal boundary up to 6,500 feet above ground *(see fig. 35 and photo, page 9)*.

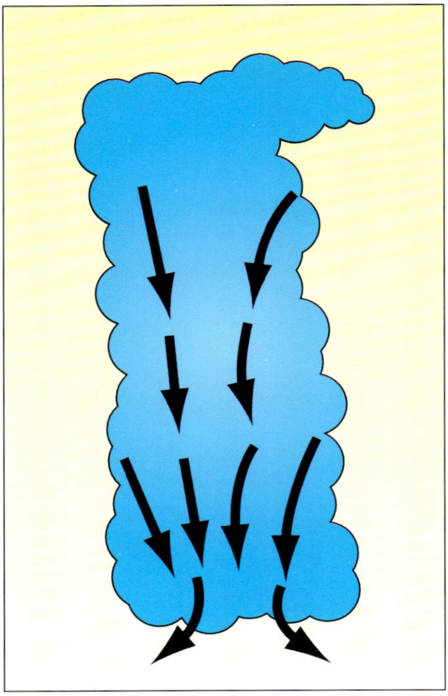

Fig. 34 — Dissipating stage

(b) Downburst and Microburst

If the downdraft underneath a cell is exceptionally severe so that damaging winds result, it is referred to as a "downburst." Such occurrences happen almost every summer in British Columbia but only rarely hit towns and are usually only evidenced by the blowdowns in the forests. On occasion, embedded with the downburst are exceptionally violent winds that are about a mile across and last for only a few minutes. These winds are known as "microbursts" and can literally force an aircraft into the ground.

(c) Funnel Cloud, Tornado and Waterspout

The most violent thunderstorms draw air into their base with great vigor. The incoming air tends to have some rotating motion, and, if it should become concentrated in a small area, forms a rotating vortex in the cloud base in which wind speeds can exceed 200 knots. If the vortex becomes strong enough, it will begin to extend a funnel-shaped cloud downwards from the base. If the cloud does not reach the ground it is called a funnel cloud, if it reaches the ground, it is a tornado; if it touches water, it is a waterspout. All are extremely hazardous to aircraft. Fortunately, the mountainous terrain does not readily allow the required rotation to develop, making tornado events rare. The area in British Columbia with the highest probability of funnel clouds or tornadoes is over the plateau area of the central interior.

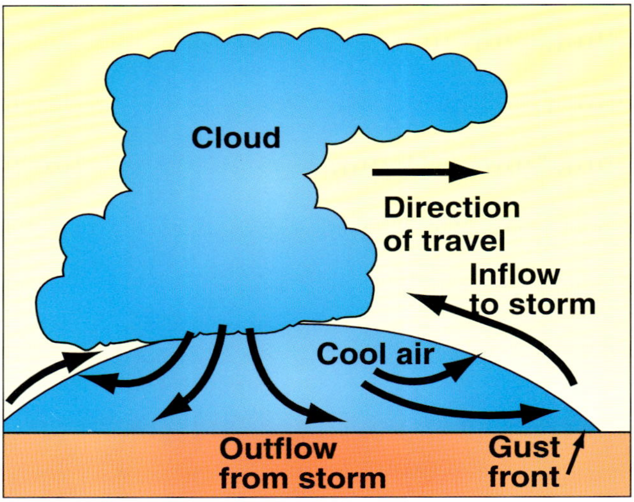

Fig. 35 — The gust front

Waterspouts tend to be a coastal phenomenon and do occur more often than tornadoes because of the frequency of which cold air is projected over the warm water, enhancing the instability. The first sign that a waterspout may form is if the cloud sags down in one area. If this bulge continues downward to the sea surface, forming a vortex beneath it, sea water will be carried aloft in the lower 60 or 100 feet.

Types of Thunderstorms

Thunderstorms are usually named in such a manner as to describe the way they were formed.

(a) Air Mass Thunderstorms

These thunderstorms form within a warm, moist air mass and are a product of solar heating. They are normally isolated, are seldom violent and usually only occur during the time of peak heating *(see fig. 36)*.

Fig. 36 — Air mass thunderstorm

(b) Frontal Thunderstorms

These thunderstorms form either as the result of a frontal surface lifting an unstable air mass or a stable airmass becoming unstable as a result of the lifting. Thunderstorms tend to be numerous in the area, are often embedded in other cloud layers, and tend to be active during the afternoon and well into the evening. Cold frontal thunderstorms are normally more severe than warm frontal thunderstorms *(see fig. 37)*.

(d) Squall Line Thunderstorms

A squall line is a line of thunderstorms that tends to develop 100 to 200 miles ahead of a fast-moving cold front. These lines tend to form rapidly, have lower bases and higher tops than most thunderstorms and are similar but more violent than cold front thunderstorms. While these storms can occur at anytime, they tend to be rare in British Columbia and the Yukon.

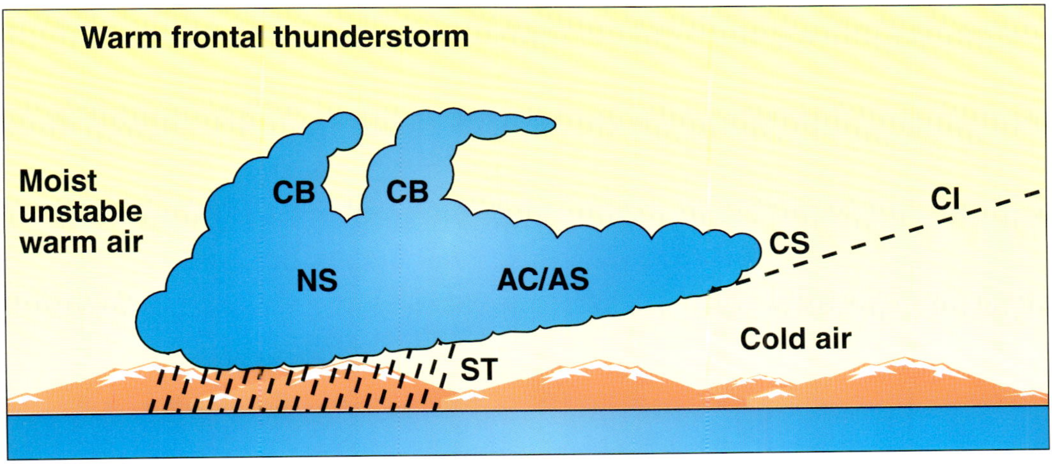

Fig. 37 — Frontal thunderstorms

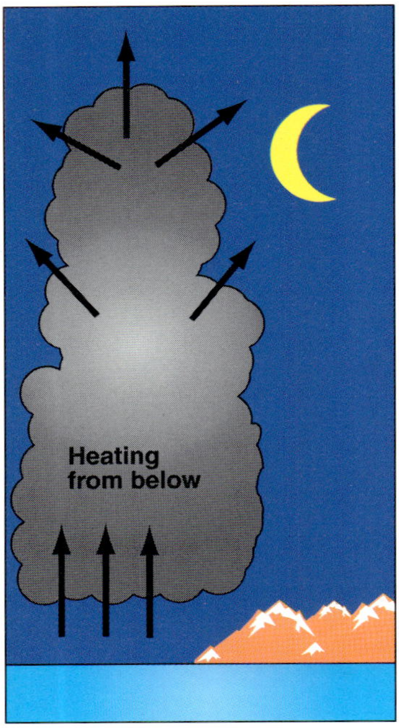

Fig. 38 — Convective thunderstorms

(e) Convective Thunderstorms

Convective thunderstorms are isolated storms created by heating from below. The most common type occur at any time of day along the British Columbia coast during the fall, winter and spring when cold air moves out over the water *(see fig. 38)*.

(f) Orographic Thunderstorms

Orographic thunderstorms occur when semi-moist, unstable air is forced up a mountain slope. The amount of lift required varies with the amount of moisture present in the air. This type of thunderstorm is most common during the afternoon and early evening, and is usually isolated but on occasion will form a long, unbroken line along a mountain slope. More violent orographic thunderstorms with hail are fairly common along the Rocky Mountains *(see fig. 39)*.

(g) Nocturnal Thunderstorms

Nocturnal thunderstorms are those that develop during or persist all night. Usually, they are associated with an upper level weather feature moving through the area, are generally isolated, and tend to produce a lot of lightning.

Fig. 39 — Orographic thunderstorms

Fronts

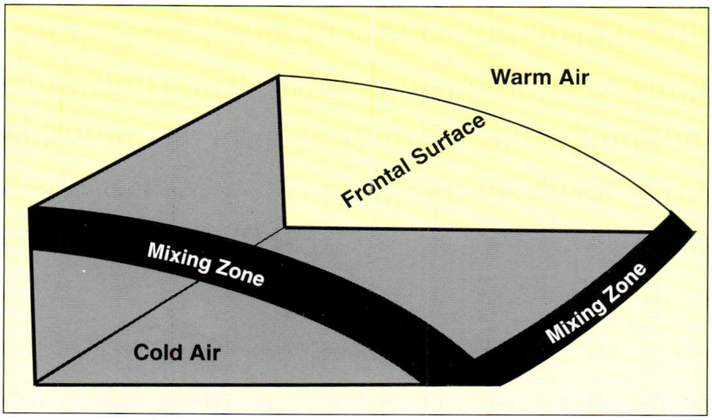

Fig. 40 — Cross-section of a cold front

A frontal zone is the transition zone between two air masses. If the colder air mass is advancing, then the leading edge of the frontal zone is described as being a cold front. If the colder air mass is retreating, then the trailing edge of the frontal zone is described as being a warm front. While only the surface front is shown on a weather map, it is important to realize that an air mass is three-dimensional and resembles an "inverted saucer" *(see fig. 40)*.

The movement of a front is dependent on the motion of the cold air perpendicular to the front, both at the surface and aloft. When the winds blow across the front, it tends to move with the wind. When winds blow parallel to the front, the front moves slowly or even becomes quasistationary. The motion of the warm air does not affect the motion of the front.

Finally, on surface charts, fronts are usually drawn as relatively straight lines. In reality, this is seldom so. Cold air flows across the surface like water. When advancing, it readily moves across level ground but in mountainous terrain it is held up until it either finds a gap or deepens to the point where it can move over the barrier. Cold air also readily accelerates downhill resulting in rapid motion along valleys such as the Fraser Canyon and the Rocky Mountain Trench. When retreating, cold air moves slowly and leaves pools of cold air in low lying areas that take time to modify out of existence.

Frontal Weather

When two different air masses encounter each other across a front, the warm air will be lifted by the cooler, denser air. When this happens, the weather at a front can vary from clear skies to widespread cloud and rain with embedded thunderstorms. The weather occurring at a front depends on:

(a) the amount of moisture available

Sufficient moisture must be present for clouds to form. Insufficient moisture results in a "dry" or "inactive" front that may be marked by only changes of temperature, pressure and wind. An inactive front can become active quickly if it encounters an area of moisture.

(b) the stability of the air being lifted

The degree of stability influences the type of cloud being formed. Unstable air will produce cumuliform clouds accompanied by showery weather and more turbulent conditions. Stable air will produce stratiform cloud accompanied by steady precipitation and little or no turbulence.

(c) the slope of the front

A shallow frontal surface such as a warm front produces widespread cloud and steady precipitation. Such areas are susceptible to the formation of low stratus cloud and fog and may have an area of freezing precipitation. Passage of the front is usually noted by the end of the steady precipitation followed by a slow reduction in the cloud cover.

A steep frontal surface such as a cold front tends to produce a narrow band of convective weather. Although blustery, the period of bad weather is short-lived and the improvement behind the front is dramatic *(see fig 41)*.

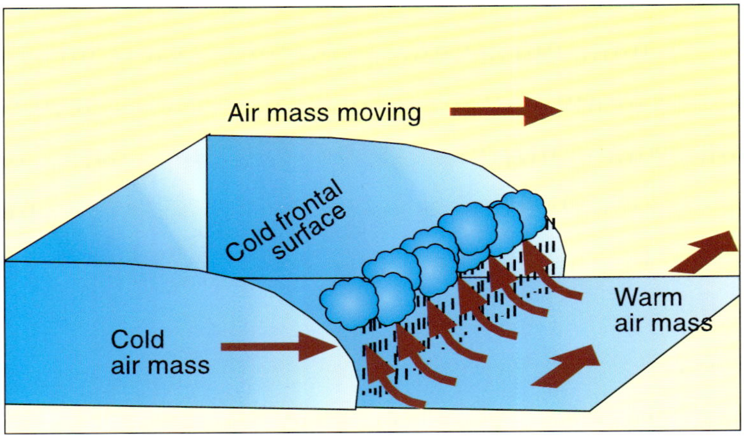

Fig. 41 — Cross-section of a cold front

(d) the speed of the front

A fast-moving cold front enhances the vertical motion along the front and in turn causes the instability to be accentuated. The result is more vigorous convective-type weather and the potential for the development of squall lines and severe weather.

Frontal Waves and Occlusions

Small-scale changes in pressure along a front can create localized alterations in the wind field resulting in bending of the front. This bending takes on a wave-like appearance as part of the front begins to move as a warm front and another part moves as a cold front. Such a structure is known as a frontal wave. There are two types of frontal waves:

(a) Stable wave

The wave structure moves along the front but does not develop beyond the wave appearance. Such features, known as stable waves, tend to move rapidly (25 to 60 knots) along the front and are accompanied by a localized area of heavier cloud and precipitation. The air mass stability around the wave determines the cloud and precipitation type. Since the wave moves rapidly, the associated weather duration tends to be short *(see fig. 42)*.

(b) Occluding Waves

Given additional support for development, such as an upper trough, the surface pressure will continue to fall near the frontal wave, causing the formation of a low pressure centre and strengthening winds. The wind behind the cold front increases causing the cold front to accelerate and begin to wrap around the low.

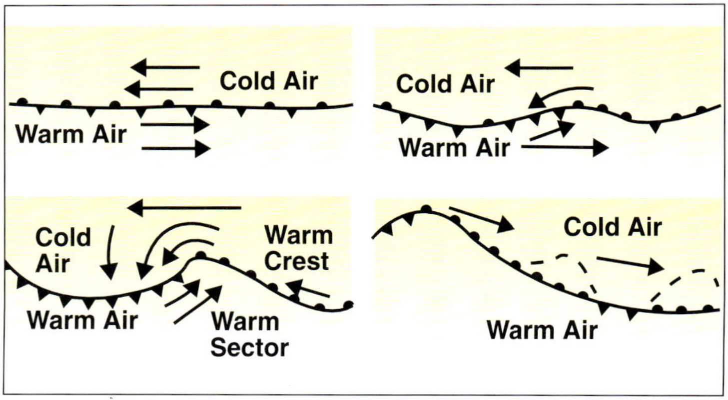

Fig. 42 — Frontal waves

Eventually, it catches up with the warm front and the two fronts occlude or "close together." At this point, the low is at maximum intensity. For British Columbia and the Yukon, this development occurs almost exclusively over the Gulf of Alaska and the Eastern Pacific *(see fig. 43)*.

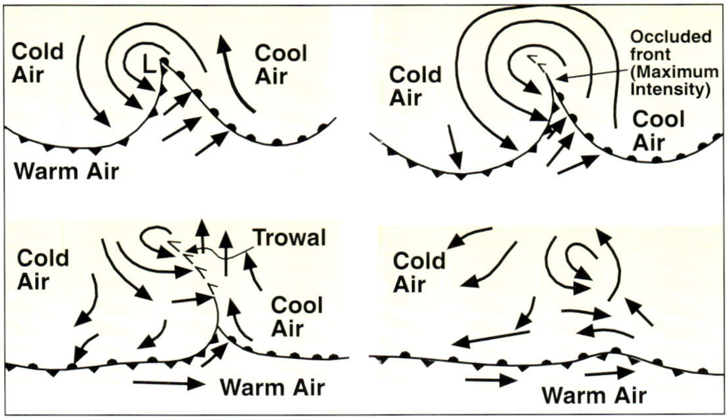

Fig. 43 — Formation of an occluding frontal wave

Occlusions occur because the air behind the cold front is colder and denser than the cool air mass ahead of the warm front. Thus it undercuts not only the warm sector of the original wave but also the warm front, forcing both features aloft. As the warm sector is lifted higher and higher, the surface portion becomes smaller and smaller. Along the occlusion, the weather is a combination of a warm front and a cold front; that is, a mix of layer clouds with steady precipitation and embedded convective clouds with enhanced showery precipitation. Such a cloud mass should be approached with caution as both icing and turbulence can be quite variable. Eventually, the frontal wave and occlusion moves away from the low leaving only an upper frontal band curling back towards the low. This upper structure continues to weaken as it moves further and further away from the low that initially formed it *(see fig. 44)*.

Fig. 44 — An occlusion, with the warm sector being lifted higher as the warm sector becomes smaller

Cold Weather Operations

Operating an aircraft in extremely cold weather conditions can bring on a unique set of potential problems.

Temperature Inversion

An intense low level inversion is common in most mountain valleys during the winter due to cold air drainage. While most lakes in the Yukon and central British Columbia freeze over, several of the larger lakes in the southern interior of British Columbia do not freeze. The effect of these open lakes is to provide a moisture source below the inversion which allows for the development of widespread valley cloud.

When extremely cold air moves out over the open lakes, especially warmer lakes like Okanagan Lake, it becomes extremely unstable. Cloud can be seen to almost be "boiling" off the waters surface and vortices of cloud can be seen rotating upwards off the lake into the cloud. Such a condition can be very turbulent and there is a significant risk of serious icing. At the same time, the lake effect snowfall is enhanced giving areas of poor visibility.

Another interesting effect in cold air is the bending of low angle light rays as they pass through the inversion. This bending creates an effect known as "looming," a form of mirage that causes objects normally beyond the horizon to appear above the horizon.

Ice Fog

Ice fog is very common in cold air and is the result of water vapour sublimating directly to ice crystals. In conditions of light winds and temperatures colder than -25°C or so, water vapour from anthropogenic sources (man made) can form widespread and persistent ice fog. In light winds, the visibility can be reduced to near zero, closing an airport for hours or even days.

Blowing Snow

Blowing snow can occur almost anywhere where dry snow can be picked up by strong winds but poses the greatest risk in northeastern British Columbia and the northern Yukon. As winds increase, the snow begins to bluster and can, in extreme conditions, reduce horizontal visibility at runway level to less than 100 feet.

Whiteout

"Whiteout" is a phenomena that occurs in the northern Yukon when a layer of cloud of uniform thickness overlays a snow or ice-covered surface. Light rays are diffused when they pass through the cloud layer so that they strike the surface from all angles. This light is then reflected back and forth between the surface and cloud eliminating all shadows. The result is a loss of depth perception, the horizon becomes impossible to discern and dark objects seem to be floating in a field of white. Disastrous accidents have occurred under such conditions where pilots have flown into the surface unaware that they were descending and confident that they could see the ground.

Altimetry Errors

The basic barometric altimeter in an aircraft assumes a standard change of temperature with height in the atmosphere and, using this fact, certain pressure readings by the altimeter have been defined as being at certain altitudes. For example, a barometric altimeter sensing a pressure of 700 millibars would assume it was located approximately ten thousand feet above sea level.

Cold air is much more dense than the assumed value used in the standard ICAO atmosphere. For this reason, any aircraft that is flying along a constant pressure surface will actually be descending as it moves into areas of colder air, although the indicated altitude will remain unchanged. Interestingly enough, a new altimeter setting obtained from a site in the cold air will not necessarily correct this problem and may increase the error.

Consider:

A pilot obtained an altimeter setting of 29.85 inches and plans to maintain a flight level of 10,000 feet enroute. As the aircraft moves into an area with a strong low level inversion and very cold surface temperatures (mountain valleys in winter) the plane, descends gradually as it follows the constant pressure surface corresponding to an indicated altitude of 10,000 feet. A new altimeter setting, say 30.85 inches, is obtained from an airport located in the bottom of a valley, deep in the cold air. This new setting is higher than the original setting and, when it is entered, the altimeter will show an increase in altitude (in this case the change is one inch and so the altimeter will show an increase from 10,000 to 11,000 feet). Unaware of what is happening, the pilot descends even further to reach the desired enroute altitude compounding the height error.

If the aircraft were operating in cloud-shrouded mountains, an extremely hazardous situation can develop. There is no simple solution to this problem other than to be aware of it and allow for additional altitude to clear obstacles.

Volcanic Ash

A major threat to aviation is volcanic ash. When a volcano erupts, a large amount of rock is pulverized into dust and blasted upwards. The altitude is determined by the severity of the blast and, at times, the ash plume will extend into the stratosphere. This ash then spreads downwind by the winds aloft in the troposphere and the stratosphere.

The dust in the troposphere settles fairly rapidly and can limit visibility over a large area. For example, when Mt. St. Helen's erupted, there was ash fallout and limited visibility across southern Alberta and Saskatchewan.

Of greater concern is the volcanic ash that is ingested by aircraft engines at flight level. Piston-driven engines have failed due to plugged air filters while turbine engines have "flamed out."

The volcanic dust also contains considerable pumice material. Leading edges such as wings, struts, and turbine blades can all be abraded to the point where replacement becomes necessary. Windscreens have been abraded until they become opaque.

Chapter 3

PATTERNS IN THE WEATHER

Thunderstorm clouds near Prince George

A. Whitman

Introduction

"Weather is what you get;
Climate is what you expect"

Weather is what happens. Weather is also transitory, seldom lasting more than a matter of hours. Climate speaks to the weather history of a location. The how and why the weather varies between seemingly identical locations. Why is Abbotsford open when all other nearby airfields are closed in fog? What are the predominate winds at Penticton? Meteorologists use their knowledge of both weather and climate when producing forecasts. This constant conflict between "what you expect" and "what you get" is unending; it is a problem that becomes much more difficult when you have to take mountainous terrain into consideration.

Geography and its Effect on Climate

Mountains are the shapers of climate. Within British Columbia and the Yukon, some

Fig. 45 — Mountains, lakes and rivers affect the weather in British Columbia

of the world's most complex interactions between the migratory weather systems and mountains occur.

Most of British Columbia is mountainous. The most significant mountain ranges are the Coast Mountains that run from Vancouver northwards to the Yukon and the Rocky Mountains which run along the border with Alberta to Jasper and then continue northwestward to just west of Fort Nelson. Between these two are a series of other mountain ranges with narrow valleys between them. West of the Coast Mountains are the Insular Mountains of Vancouver Island and the Queen Charlotte Islands.

Fig. 46 — The Yukon

Three major plateaus exist in British Columbia. These are the Central Plateau between Williams Lake and Prince George, the extension of the Canadian Prairies over northeastern B.C. and the Atlin-Stikine Plateau over northwestern British Columbia.

The Yukon is also mountainous. The greatest concentration of 16,000 foot plus mountains in North America, the St. Elias Mountains, dominate the southwest corner and extend southeastwards at lower elevations to meet the Coast Mountains. Further inland, the Cassiar Mountains extend northwards into the southern Yukon to join the Pelly Mountains over south-central Yukon. The Mackenzie-Selwyn Mountains straddle the Yukon-Northwest Territories border before turning westwards and narrowing over the Yukon north of Dawson and Mayo as the Ogilvie Mountains. The relatively low Richardson and British Mountains form another barrier near the Arctic coast and Mackenzie Delta.

Between these mountain ranges are the Yukon River, the Liard Basin and the Porcupine Basin-Peel Plateau.

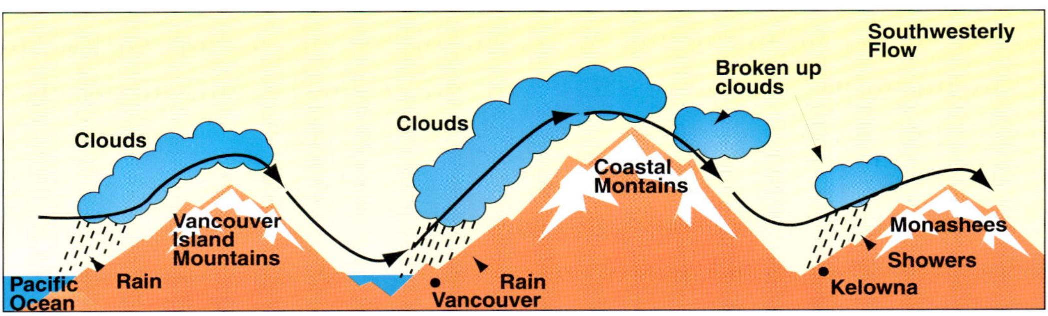

Fig. 47 — The most notable effect of mountains is their impact on precipitation

How are mountains "the shapers of climate?" On the small scale, topographical features such as elevation, orientation to the mean winds, slope, and exposure all combine to alter the local climate. Higher elevations tend to have lower average temperatures and increased precipitation. Low-lying areas such as valleys tend to allow cold air to drain into them, creating higher occurrences of frost and fog.

On the larger scale, the most noticeable effect of mountains is their impact on precipitation and temperature extremes *(see fig. 47)*. Along the windward slopes, the air is forced to lift, resulting in steady, widespread precipitation. Such an effect creates the rain forests of the west coast of British Columbia. On the leeward side, subsidence creates rain shadows where precipitation is sharply reduced. Such an effect produces the arid and semi-arid regions of the southern interior of B.C. and southwestern Yukon. Airports such as Fort St. John see Chinook conditions in winter that can quickly drive the temperature above freezing.

Mountain valleys allow cold air to pool creating inversions. Such conditions have set record cold temperatures in the Yukon, while in the southern interior of British Columbia, the inversion traps moisture near the surface supporting the development of low level cloud. The Rocky Mountains also act as a barrier. All but the strongest incursion of Arctic air are held out of the southern British Columbia interior, maintaining temperatures much above those on the Prairies. The Ogilvie Mountains also act as an early barrier preventing Arctic air from flooding over central and southern Yukon.

The Mean Atmospheric Circulation System

The mean circulation of the upper level winds over British Columbia and the Yukon is the result of two semi-permanent features. The "Pacific High" extends from the area to the west of Hawaii to the west coast of the United States and the "Aleutian Low" occurs in the Gulf of Alaska and over the Aleutian Islands. The combination of the counterclockwise circulation around the Aleutian Low and the clockwise circulation around the Pacific High

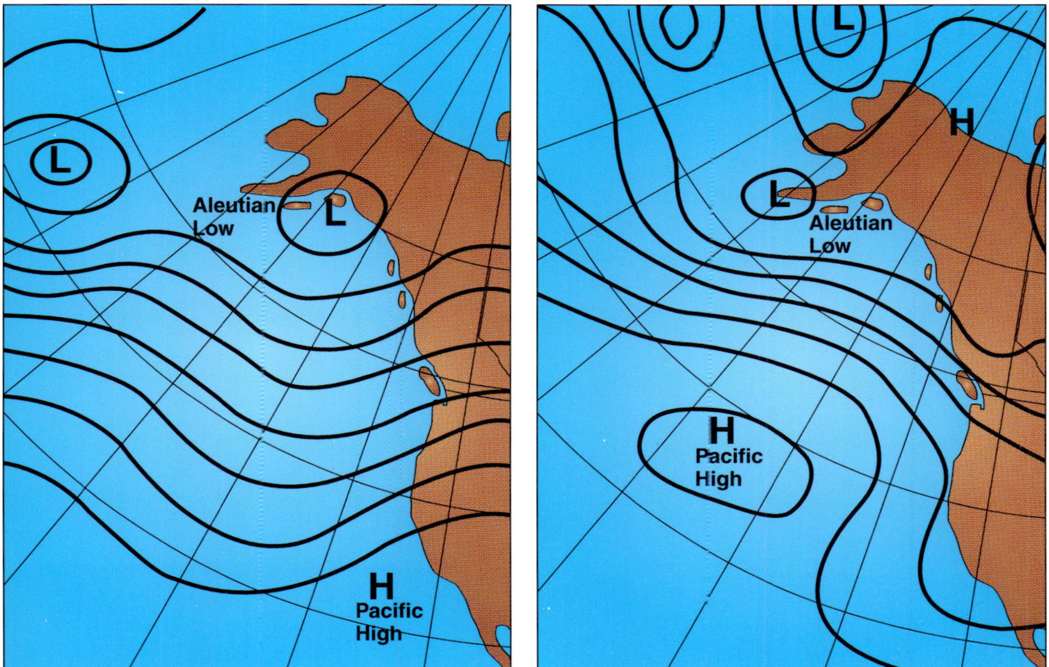

Fig. 48A — Typical winter pattern

Fig. 48B — Typical summer pattern

results in a mean westerly flow onto the west coast of North America

The Aleutian Low is strongest in the winter which causes the mean circulation to take on a southwesterly component while the Pacific High is strongest in the summer causing a more northwesterly flow. Complicating this simple relationship are the upper troughs of low pressure, upper ridges of high pressure and other weather systems that move along with the upper winds altering or disrupting the normal pattern.

Upper Troughs and Upper Ridges

The most common features that move with the upper flow are the upper ridges and upper troughs. Upper ridges are usually strongest in the summer and produce widespread areas of clear skies with little or no convective cloud. In the winter, the upper ridges are weaker and often contain considerable cloudiness. This cloud however, will produce little or no precipitation.

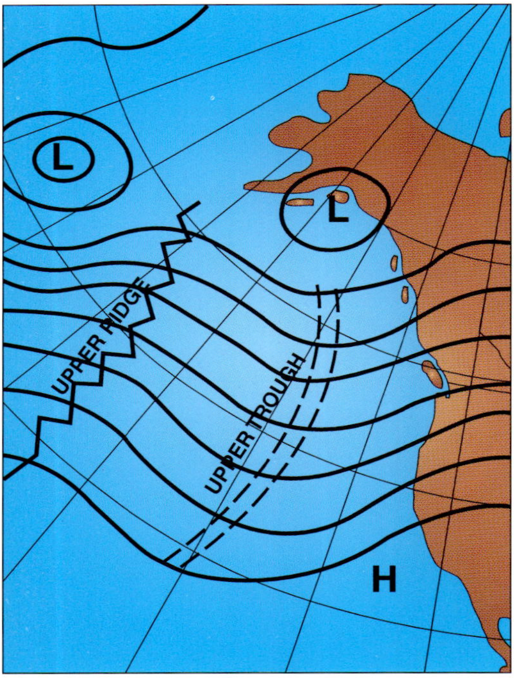

Fig. 49 — Typical winter pattern with upper troughs and ridges added

Upper troughs, because of the induced vertical lift, are areas of cloud and precipitation. Upper troughs tend to be strongest in the winter and often have broad cloud shields and widespread precipitation, particularly in upslope areas along the windward slopes of the mountain ranges. During the summer months, the cloud shields associated with upper troughs are narrower, usually quite convective and produce mainly showers and thundershowers. Upper troughs may have a surface low pressure system or a frontal system associated with them further enhancing the cloud and precipitation. Clearing behind an upper trough can be gradual but tends to be quite rapid in the summer.

Both the upper trough and the upper ridge interact with the Pacific High and the Aleutian Low. An upper ridge will strengthen the Pacific High and weaken the Aleutian Low. This tends to dissipate or divert cloud northwards away from the British Columbia coast and into Alaska or across the Yukon. An upper trough will cause the Aleutian Low to deepen and the Pacific High to weaken so that weather systems will follow a general westerly track over the Gulf of Alaska and into British Columbia.

Migratory Surface Weather Systems

Closer to the ground are the migratory or traveling surface weather systems (low pressure areas, high pressure areas, frontal systems) that are carried into British Columbia and the Yukon and produce the day-to-day weather. These traveling surface weather systems vary in intensity with the seasons and are more frequent in the winter months (Oct. to Apr.). On average, 10 to 15 such storms will occur monthly during the winter.

Winter Storms

During the winter, low pressure systems develop over the Pacific Ocean and move toward the coast. Most of these storms are either Gulf of Alaska lows, which tend to remain well offshore, or coastal lows which approach the coast before developing rapidly. Coastal lows, while not as powerful as the Gulf of Alaska low, can be quite dramatic in terms of their development and speed of movement.

On a few occasions during the winter, a low will combine the characteristics of both the Gulf of Alaska low and the coastal low. In this situation, the low first moves into the Gulf of Alaska and begins to weaken. The front sweeps over the coast giving strong winds and widespread precipitation, especially along windward slopes. After a time the low, which now sits over the Gulf of Alaska, begins to drift southeastward toward the British Columbia coast and strengthens once again.

The map shows the average tracks which the different types of lows usually follow. The actual track of any individual storm can vary somewhat from those shown.

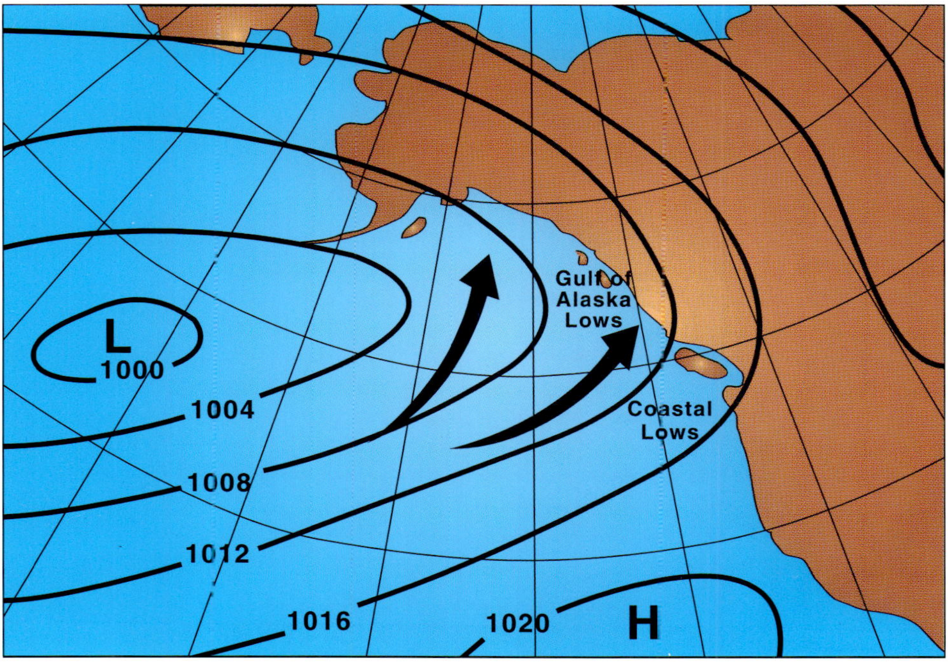

Fig. 50 — Principal winter storm tracks are superimposed on a January mean sea-level pressure pattern

(a) Gulf of Alaska Lows

A Gulf of Alaska low usually forms south of the Aleutians as a frontal wave between the cold northern air and warmer air to the south. Such a wave may travel a considerable distance eastward before it begins to take shape as a low pressure system. Once the low starts to develop, pressures fall rapidly and the entire low pressure system increases in size.

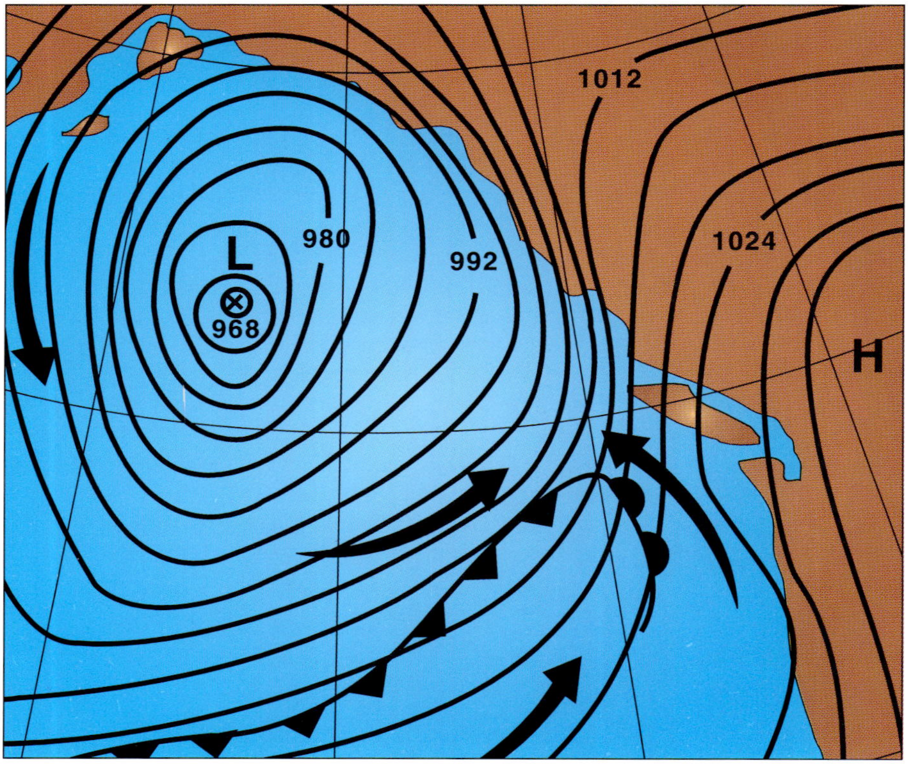

Fig. 51 — This sea-level pressure pattern indicates a Gulf of Alaska low with the associated frontal system approaching the B.C. coast

The low traveling eastwards at typically 35 to 40 knots reaches its lowest central pressure (970 millibars or lower) over the Gulf of Alaska then turns northeastward toward the northern end of the Alaskan Panhandle. The frontal system that accompanies the low continues eastward onto the coast, bringing widespread cloud, precipitation, and strong winds. Behind the cold front, a period of strong northwesterly winds of 35 to 50 knots heralds the arrival of a colder, unstable airmass.

(b) Coastal Lows

Coastal lows usually intensify very rapidly just before they move over the British Columbia coastal waters and can change from a very weak system into a severe storm in as little as 9 hours. Lows which do develop in such a rapid, or explosive manner, are referred to by the forecasters as "bombs." On the coast, very strong winds will occur usually to the east and southeast of the low, just ahead of the associated frontal system. Winds here may reach southeasterly 70 knots with gusts to 100 knots in the most severe storms. Often a second band of strong winds occur behind the cold front in the area to the southwest of the low pressure centre. Here, winds may range up to 65 knots from the west or northwest. Once the low moves ashore, it will fill rapidly over the Coast Mountains and

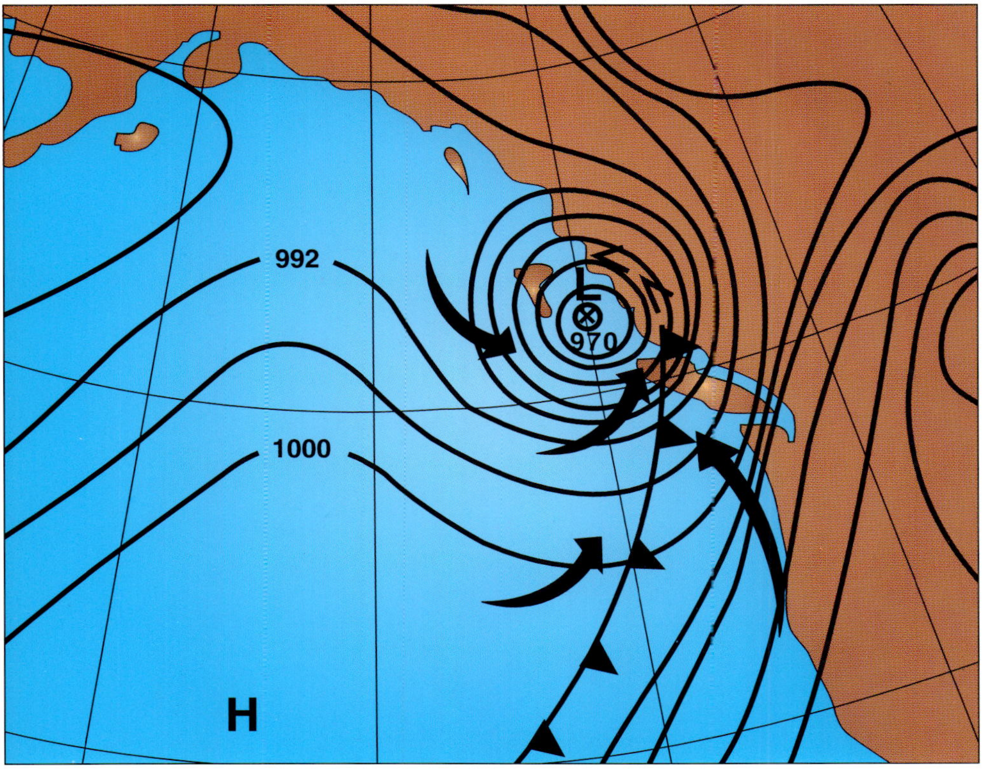

Fig. 52 — A typical sea-level pressure pattern for a coastal low and associated frontal system with wind pattern superimposed

frequently dissipates before penetrating very far into the interior.

The coastal lows often move through Queen Charlotte Sound or over the Queen Charlotte Islands. Occasionally, a low will move eastward, passing just south of Vancouver Island. The lows which follow this southern track can bring the strongest winds and heavy rains to the Vancouver and Victoria areas.

(c) Beaufort Sea Lows

A secondary storm track across the Bering Sea then eastwards across the Beaufort Sea, active mainly during the fall and the winter will produce mild temperatures with strong southwest winds through the British and Richardson Mountains. This is usually accompanied by bitterly cold air over the southern interior of the Yukon.

(d) Winter frontal systems

While low pressure systems either die along the Coastal Mountains or curl northward into the Gulf of Alaska, the associated frontal systems will push across the coast and inland. The favourite track takes the frontal wave across central British Columbia with the trailing cold front sliding southward across the south coast and southern interior of British Columbia. An occluded front usually extends northward from the frontal wave and moves across northern British Columbia and into the southern Yukon. Penetration into the Yukon is usually limited to the southern areas as the northern areas are sheltered under an Arctic High.

Along the coast, the weather tends to be quite dramatic as the approaching warm front spreads extensive, deep-layered cloud, steady precipitation and strong southeast winds across the area. With the passage of the warm front, the precipitation becomes intermit-

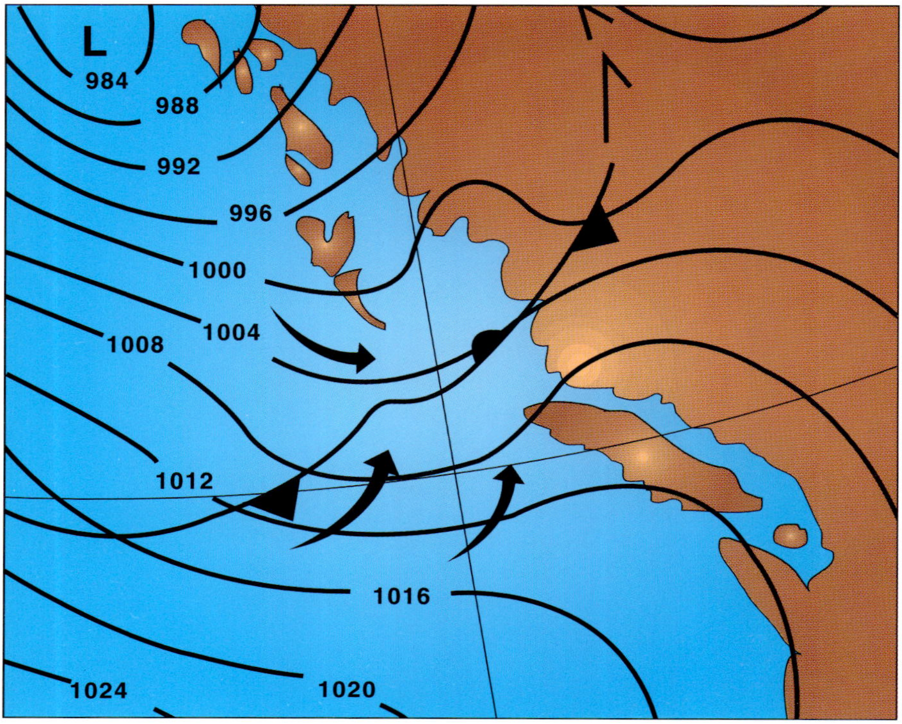

Fig. 53 — A typical winter pressure pattern shows a front crossing the coast with an indication of the winds near the front

tent, the lowest cloud layers break up but the higher cloud layers persist. At the same time, strong southerly winds usually persist. The arrival of the trailing cold front is marked by the development of showery conditions which can be both widespread and heavy in intensity. In its wake, strong gusty northwesterly winds often prevail. With the most active fronts, warm or cold, the winds can briefly rise to 60 knots with gusts above 80 knots at locations over the northern end of Vancouver Island and around the Queen Charlotte Islands.

As the front moves inland, it tends to weaken due to subsidence to the lee of the Coastal Mountains and the St. Elias Mountains but still gives steady or intermittent precipitation whose type will vary with the local temperatures. Accumulations are usually light in comparison to the coast. One notable exception is along the Rocky Mountains between Revelstoke and Blue River where upslope lift often produces much heavier precipitation. Similar effects can be seen along the slopes of the Ogilvie Mountains and Mackenzie Mountains in the Yukon.

Air mass thunderstorms occur most often during the winter in the very cold air that follows the passage of a cold front. They often form into squall lines which move toward the coast about 12 to 24 hours after the frontal passage.

(e) Winter high pressure systems

Surface high pressure systems, long valued for their generally good weather, are stronger in the winter than in the summer. Along the coast, it is the ridges of high pressure that provide the only break between active weather systems. As the ridge of high pressure approaches the coast the higher level clouds dissipate and the lower cloud layer break up leaving widespread convective cloud. The convective cloud most often consists of cumulus and towering cumulus clouds and gives showers especially along the slopes of mountains. When the contrast between the temperature of the air and water is great

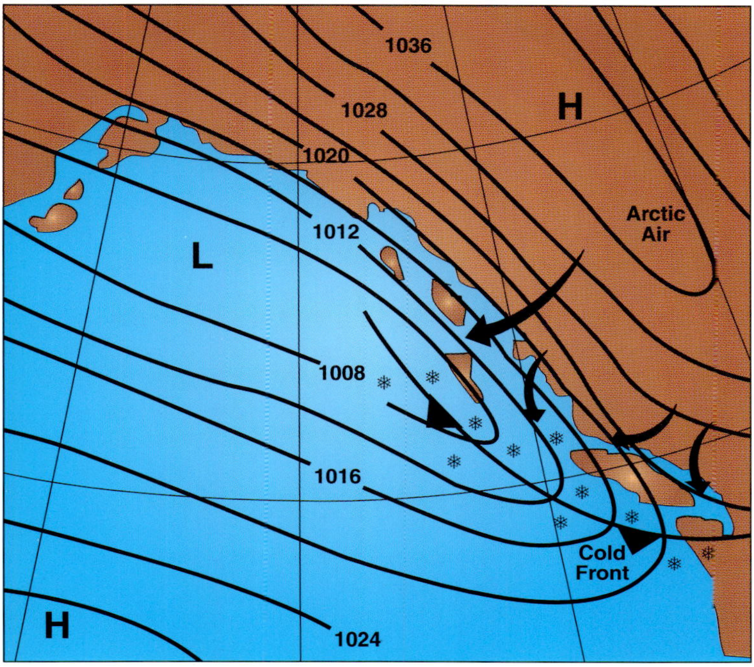

Fig. 54 — A ridge of high pressure builds over the province as the cold, arctic air flows into the Interior. This pushes the cold front out onto the coast. Strong outflow winds occur through the mainland inlets and near the mouth of the inlets

enough, baby cumulonimbus will develop giving much heavier showers, snow pellets or small hail, and even localized thunderstorms.

In the southern interior of British Columbia, areas of high pressure have a lesser impact because of the widespread valley cloud. Cold air stagnating in the bottom of valleys causes a strong low level inversion to form which traps any moisture from local sources. This moisture eventually forms cloud which, despite being only a few thousand feet thick, fills entire valley systems and persists for weeks. This cloud is usually dry and will only produce light snow in areas where the moisture supply has been increased. Valley cloud will only move if strong winds develop in the valley or it will dissipate if the major moisture sources freeze over. Since the larger lakes in the southern interior of British Columbia freeze over completely only in the coldest winters, most southern valleys remain susceptible to the development of valley cloud from December to mid-February. Passing ridges of high pressure have little impact on valley cloud.

The northern half of British Columbia and the Yukon is subject to widespread valley cloud only during the early part of the season, as the lakes and rivers generally freeze over completely. Thus, ridges of high pressure during mid and late winter bring widespread clear, cold weather.

(f) Arctic Outbreaks

During winter, a strong area of high pressure forms in the very cold air over Alaska, the Yukon and the northern end of the Mackenzie River valley. This cold arctic air moves southeastwards into the Prairies but can also spread over northern and central British Columbia. Most often, the Arctic air pushes southward into the central interior before coming to rest. At the same time, Arctic air also flows through the mountain passes from Alberta and fills the Rocky Mountain Trench. Depending on the strength of the arctic front, winds can shift abruptly into the northwest with the passage of the front and be gusty for several hours. At least once or twice each year, the advance of Arctic air is so strong in British Columbia that it spreads into the southern interior

Arctic air over the interior offers little problems other than the temperature. Most of the valley cloud dissipates giving clear cold days and nights. Over the bodies of water that have remained unfrozen, such as Okanagan Lake, sea smoke will form over the water and lift to form cumulus clouds. These clouds, if there is a significant difference between the air temperature and the water temperature, will be turbulent, contain icing and produce local heavy snowshowers. Other than this, VFR conditions will prevail except for some localized problems with ice fog.

(g) Outflow

If the cold air deepens sufficiently over the interior of British Columbia, it can flow through the coastal mountains, down the coastal inlets, and cascade out over the coastal waters far enough to cover the Queen Charlottes and Vancouver Island. This condition of cold air spilling through the coastal inlets is referred to as "outflow" and can persist without respite, for days. Outflow is a common occurrence over the north and central coast areas of British Columbia and infrequent over the south coast.

Along the mainland coast, a band of cloud and showers or flurries will accompany the leading edge of the outflow, to be followed by clearing skies. These clear skies will then persist throughout the period of outflow winds. Offshore, the cold dry air flowing over bodies of water such as Georgia Strait and Hecate Strait will become unstable and will pick up enough moisture to give heavy snowshowers along the east side of Vancouver Island and the Queen Charlotte Islands.

The strong winds funnel down the mainland inlets often reaching speeds up to 60 knots and occasionally rising as high as 100 knots. Side tributaries from the main inlets also have strong winds and, where a major side valley joins the main inlet, chaotic conditions are found. When the strong winds flow out from the mouth of the inlets, they continue for some distance but gradually weaken as the flow is no longer confined. Extreme caution is

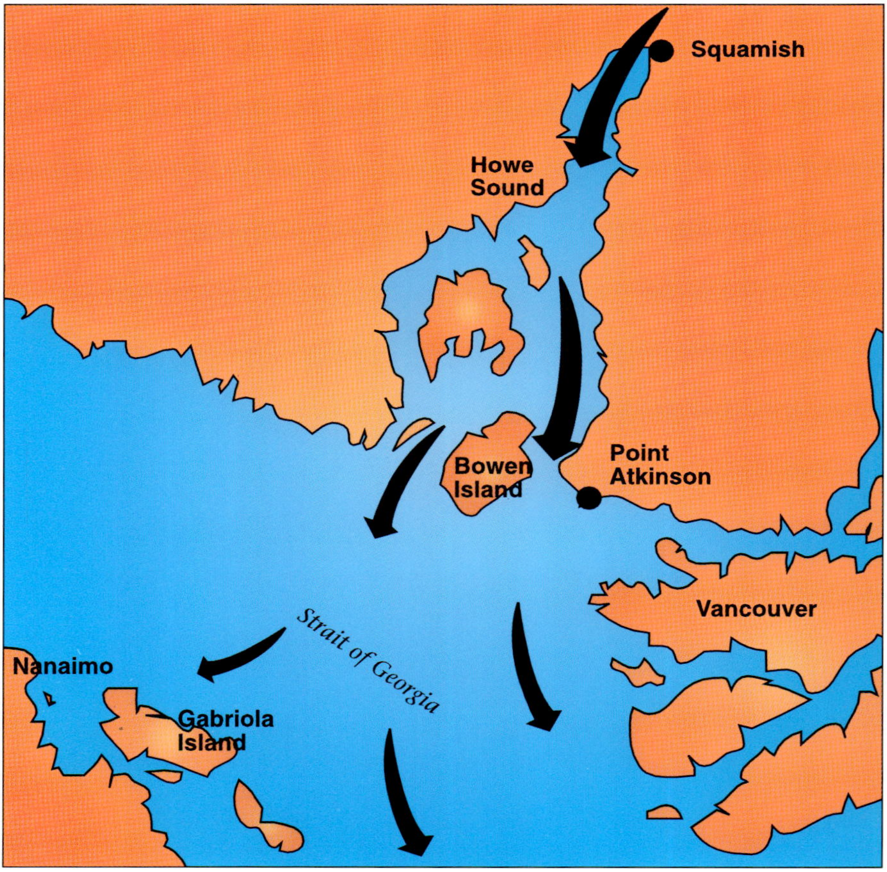

Fig. 55 — Strong winds funnel down the mainland inlets often reaching speeds up to 60 knots

advised when crossing coastal inlets during an outflow situation. The winds could increase very abruptly in a narrow band near the mouth of the inlet, especially over the North and Central Coast where the narrow inlets produce extremely strong outflow conditions.

The end of an arctic outbreak occurs when the cold air is forced back inland by the arrival of warmer air being driven ahead of a Pacific storm. The precipitation with the system may start as snow but changes to either rain or rain and snow mixed as temperatures moderate. In the inlets, however, the cold outflow winds will gradually ease as the southeast winds strengthen. Often rain will be reported at the mouth of the inlets while snow continues further up in the inlet. Freezing rain can also occur in the inlets until the cold air is fully scoured out by the approaching warmer air. The Arctic High over the Yukon can result in strong outflow winds through the coastal passes to the Alaska Panhandle and also through the Richardson and British Mountain passes, often in combination with a Beaufort Sea low. High windchills and blowing snow usually result with blizzard conditions being common.

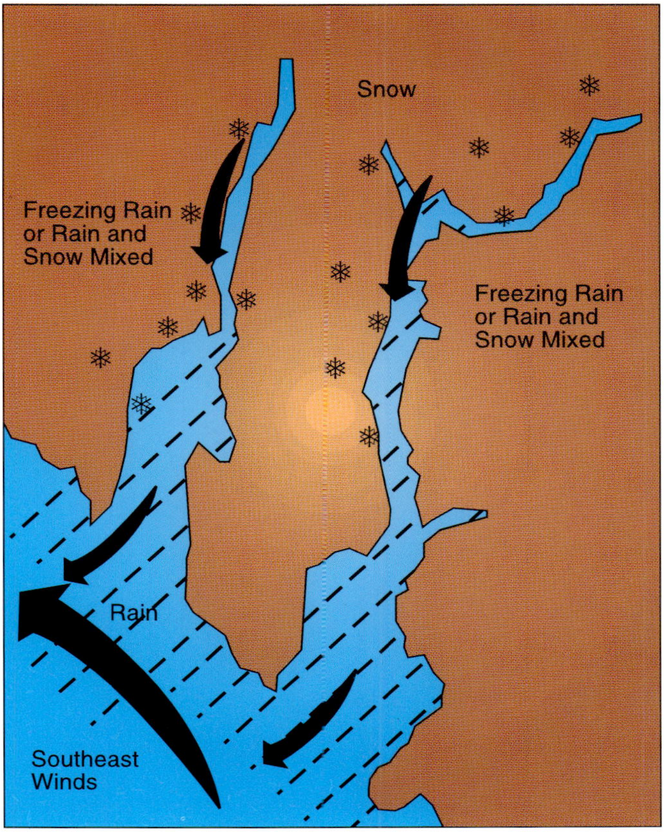

Fig. 56 — The end of an Arctic outbreak occurs when the cold air is forced back inland by the arrival of warmer air being driven ahead of a Pacific storm

Summer Weather

In the summer months (May to September), the frequency and severity of the storms is much reduced. Low pressure areas usually remain offshore as the Pacific High strengthens and moves further north. This northward shift causes the main storm track to shift into the northern Gulf of Alaska and across northern British Columbia or into the Yukon. South of this track, minor frontal systems, upper troughs and thunderstorms produce most of the weather. In August and September, weeks can pass between weather systems.

(a) Summer fronts

During the summer months fronts tend to approach the coast from the northwest across the Gulf of Alaska. Over the northern coastal areas and the southern Yukon, the front is usually accompanied by a narrow band of cloud and light rainfall. Winds will tend to increase along the coast just ahead of the front and can be expected to reach values of 30 to 40 knots over some exposed locations around the Queen Charlottes and the north end of Vancouver Island. As the front continues southward and presses further into the Pacific High, the rain area often disappears and the clouds begin to dissipate.

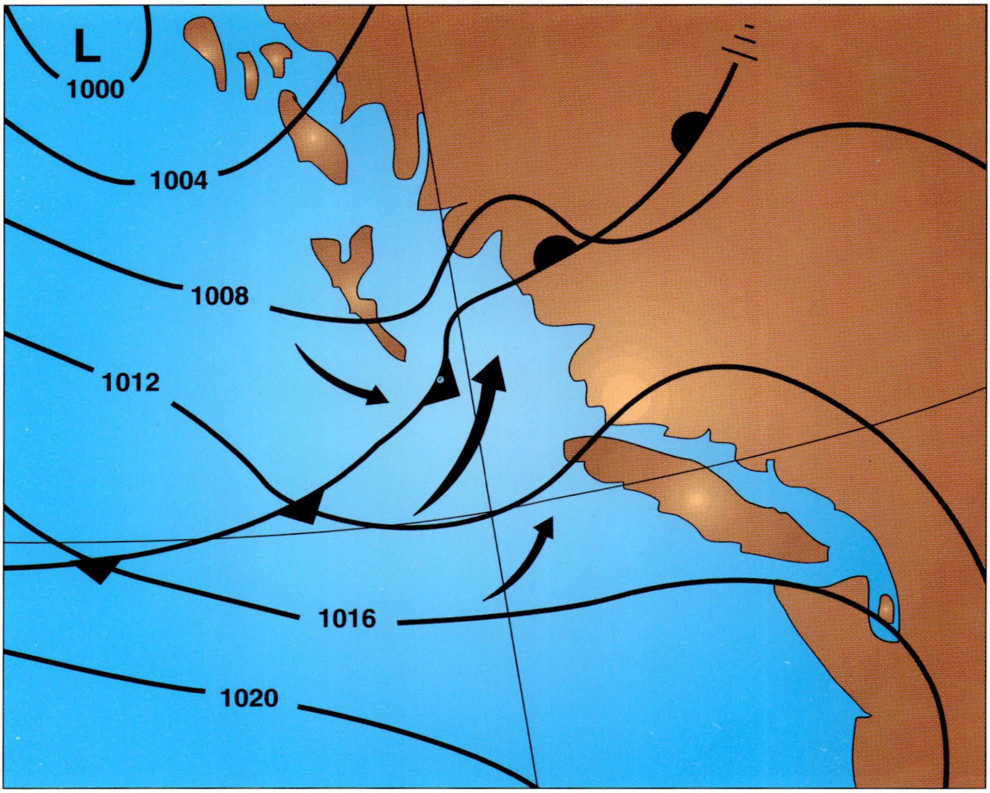

Fig. 57 — A typical summer pressure pattern shows a front crossing the coast with an indication of the winds near the front

Behind the front, pressure rises are fairly strong as the following ridge of high pressure builds towards the coast. This produces strong northwest winds following the frontal passage. The strongest northwesterly winds are often reported where the airstream is funneled between the mountains of the mainland and Vancouver Island. This effect is particularly noticeable in the spring months as the fronts still retain some of the strength of winter storms.

As the front moves inland, it is weakened by subsidence to the lee of the Coast Mountains. The cloud cover will develop noticeable breaks and precipitation, if any, will be light. In crossing the Coast Mountains, these fronts often lose their definition or disappear completely, before reappearing as a band of afternoon or evening thundershowers. These thunderstorms often persist well into the night. Further downstream, the frontal cloud band will often thicken and precipitation intensify along the upslope areas of the Rocky Mountains in British Columbia and the Mackenzie Mountains in the Yukon.

(b) Thermal Troughs

The usual summer pressure pattern has a high pressure area over the eastern Pacific and troughs of low pressure in the southern interior of B.C. and the central Yukon. This trough forms due to prolonged heating and is referred to as a thermal trough. The effect of this trough is to create a light, disorganized wind pattern as air flows towards the hottest locations. Along the coast, light winds in the morning are replaced by strengthening inflow winds during the afternoon and evening in most inlets and valleys as the cool coastal air is drawn towards the interior. Locations in B.C. with marked inflow winds are the Strait of Juan de Fuca, Portland Inlet, Howe Sound and through the Hope area and up the Fraser Canyon. Strong inflow winds developing through the coastal passes may

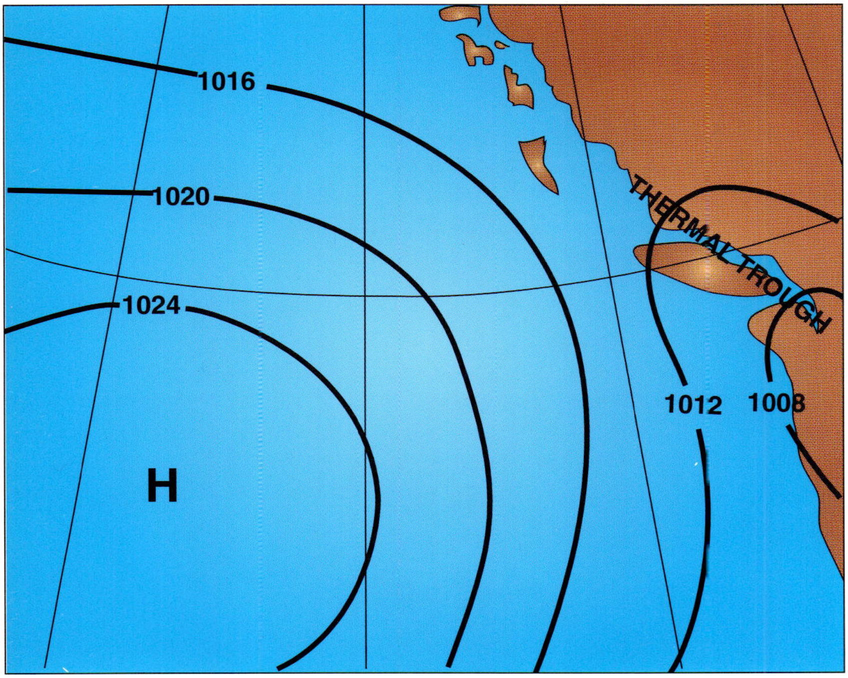

Fig. 58 — Thermal troughs

also produce strong afternoon gusts in the southwest Yukon.

Sea fog and marine stratus are prevalent over the offshore waters and usually present in the western entrance of Juan de Fuca Strait during the late summer and fall.

When inflow conditions occur, the fog and stratus is frequently drawn through the strait but usually only as far as the Victoria area. Since inflow is common, fog is frequently found in Juan de Fuca Strait for much of the year.

On occasion during the summer, the thermal trough will move out from the interior and onto the coast. When it does, it normally lies over Georgia Strait causing subsident outflow winds from the interior which gives clear skies and light winds along the coast.

After a few days, the thermals trough will shift back into the interior causing westerly winds of 20 to 30 knots through Juan de Fuca Strait and into southern Georgia Strait. Sea fog and stratus will accompany these winds and may extend as far as Vancouver Airport and Boundary Bay. In the strongest cases, the fog will lift into stratus and spread up the Fraser Valley to Hope.

(c) Thunderstorms

Although not frequent, thunderstorms do occur along the coast of British Columbia in the summer. Airmass thunderstorms are the most common, tend to develop during the late afternoon or evening and drift eastward along the sides of inlets or valleys. Although short-lived, they can produce intense hail and lightning. On rarer occasions, frontal thunderstorms will move into the coastal areas. One favorite track involves upper fronts moving northward through Washington. Thunderstorms are rare on the north coast of the Yukon.

Unlike the coast, thunderstorms are more frequent in the interior and the Yukon. While airmass thunderstorms are the most common, all other types of thunderstorms such as frontal thunderstorms and nocturnal thunderstorms will occur. The intensity of thunderstorms is also highly variable. While a high percentage will be of normal intensity, there will be several occurrences each summer of storms that reach severe intensity and provide a full range of severe weather up to and including tornadoes.

Over the southern interior of British Columbia and the Yukon, the mountainous terrain hinders the full development of thunderstorms. Over most regions, the predominant types are airmass and nocturnal thunderstorms and, when they do occur, they generally move along the valleys. Severe weather is uncommon and, when it does occur, it usually consists of severe lightning, large hail and downburst winds. While thunderstorms can occur at anytime during the summer, the main period is from June to August. The peak window in the Yukon is from mid-June to mid-August with severe weather being rare and usually limited to heavy rainfalls causing flash flooding, although dry thunderstorms are responsible for some of the worst wildfires.

In the central interior of British Columbia and the northeastern section of British Columbia, the plateau type terrain allows the thunderstorms to reach full intensity. While airmass thunderstorms still remain the predominate type, frontal thunderstorms and nocturnal thunderstorms are common. The typical scenario would see the beginning of thunderstorm activity early in the afternoon and for it to persist well into the night. Most often the thunderstorms move towards the northeast and given the right conditions, their intensity can reach the severe level. The normal thunderstorm season for both areas is June to August.

Cold Lows

A cold low is a large, nearly circular area of the atmosphere in which temperatures get colder towards the centre both at the surface and aloft. While a surface low pressure centre is usually present beneath the cold low, its true character is most evident on upper charts. The significance of cold lows is that they produce large areas of cloud and precipitation, tend to persist in one location for prolonged periods of time and are difficult to predict.

Cold lows can occur at any time of the year but the most frequent occurrence, "cold low season," is from the end of May to mid-July. At this time, pools of cold air break away from the Aleutian Low and move southeastwards to take up a nearly stationary position off the British Columbia or Washington coast. Once established, the cold low will rotate a series of upper troughs across southern British Columbia. The overall effect is to produce a widespread area of cool, unstable air in which bands of cloud, showers and thundershowers occur. Along the deformation zone to the northeast of the cold low, the enhanced vertical lift will thicken the cloud cover and produce widespread steady precipitation. In many cases, the deformation zone is where widespread and prolonged thunderstorm activity occurs.

Eventually, a strong system will approach from the west and will have sufficient strength to force the cold low inland, usually in the form of a strong upper trough. A favorite track is across the northern United States but alternate tracks are across southern British Columbia or even northeastward along a line from Seattle to Fort St. John. As it crosses the area, widespread cloud, showers, thundershowers, or even steady rain can occur for a period of 24 to 48 hours. Normally, while the original cold low is moving off, the next one is already moving across the Gulf of Alaska on its way to take up residence off the coast.

A similar but weaker feature may develop over the northern Gulf of Alaska, Alaska and the Yukon during June and July and persist for many days, producing cool, showery conditions.

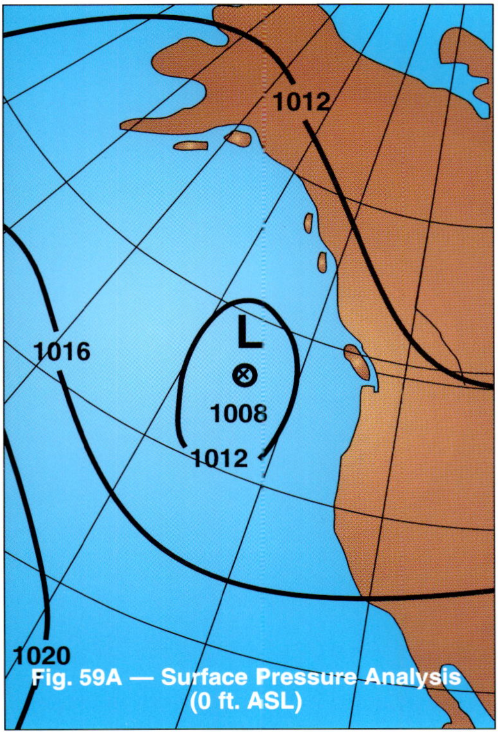

Fig. 59A — Surface Pressure Analysis
(0 ft. ASL)

Fig. 59B — 850 Millibar Analysis
(5,000 ft. ASL)

Fig. 59C — 700 Millibar Analysis
(10,000 ft. ASL)

Fig. 59D — 500 Millibar Analysis
(18,000 ft. ASL)

Chapter 4

LOCAL WEATHER HAZARDS

Entrance Lightstation, Strait of Georgia

O. Lange

MAP INDEX

Various VFR flying routes or areas are represented by circles corresponding to the page number where information on that area may be found.

Introduction

The remainder of this manual is devoted to local weather hazards experienced in B.C. and the Yukon. The focus is on the common VFR flying routes where many local aviation weather hazards are described. Even on these few specific routes, it is impossible to list all potential weather hazards. Through discussions with experienced pilots, the most common hazards are listed.

Most weather hazards are described in symbols on the many maps while others are better described in words. Table 1 provides a legend for the various symbols used throughout the local weather sections. For convenience, this table is replicated inside the back cover of this book. The maps are designed to resemble the standard VFR Navigation Charts. This should enable easy cross-referencing from the manual to VFR Navigation Charts.

On the previous page appears a quick index for this section of the manual. On a map of B.C. and part of the Yukon, various VFR flying routes or areas are represented by circles corresponding to the page number where information on that area may be found.

Table 1: Symbols Used in This Manual

Fog Symbol (3 horizontal lines)
This standard symbol for fog indicates areas where fog is frequently observed.

Cloud areas and cloud edges
Hatched lines show areas where low cloud (preventing VFR flying) is known to occur frequently. In many cases, this hazard may not be detected at any nearby airports.

Icing symbol (2 vertical lines through a half circle)
This standard symbol for icing indicates areas where severe icing is relatively common in cloud.

Choppy water symbol (symbol with two wavelike points)
For floatplane operation, this symbol is used to denote areas where winds and significant waves can make floatplane operation dangerous or impossible.

Turbulence symbol
This standard symbol for turbulence is also used to indicate areas known for significant windshear as well as potentially hazardous downdrafts.

Strong wind symbol (straight arrow)
This arrow is used to show areas prone to very strong winds and also indicates the typical direction of these winds. Where these winds encounter changing topography (hills, valley bends, coastlines, islands, etc.), turbulence, although not always indicated, can be expected.

Funneling / Channeling symbol (narrowing arrow)
This symbol is similar to the strong wind symbol except that the winds are constricted or channeled by topography. In this case, winds in the narrow portion could be very strong while surrounding locations receive much lighter winds.

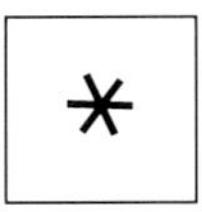

Snow symbol (asterisk)
This standard symbol for snow shows areas prone to very heavy snowfall.

Thunderstorm symbol (half circle with anvil top)
This standard symbol for cumulonimbus (CB) cloud is used to denote areas prone to thunderstorm activity. Although thunderstorms can and do occur throughout B.C. and the Yukon, these areas are noted for their thunderstorm activity.

Mill symbol (smokestack)
This symbol shows areas where major industrial activity can impact on aviation weather. The industrial activity usually results in more frequent low cloud and fog.

Mountain pass symbol (side-by-side arcs)
This symbol is used on aviation charts to indicate mountain passes, the highest point along a route. Although not a weather phenomenon, many passes are shown as they are often prone to hazardous aviation weather.

Map 1 — Vancouver to Victoria / Nanaimo

B.C. Coast — Victoria to Stewart

Outflow winds occur periodically in winter when a cold Arctic airmass covers the interior of B.C. These outflow winds can reach speeds of greater that 60 knots and can be recognized from above by observing the behavior of the water surface. Outflow winds occur in most valleys and inlets on the coast. Many of these valleys also experience inflow winds in the summer with a hot dry airmass over the interior. Inflow winds, in general, are not as strong as outflow winds. Windshear can be significant over the Strait of Georgia when strong southwesterlies blow across Vancouver Island, while southeasterlies persist in the lowest few hundred feet. Although shear is significant, this phenomena produces only minor turbulence due to the stability of the pattern (cooler air below warmer air).

Fog can be a problem year round, however a general rule usually applies. When the water temperature exceeds the air temperature (autumn and winter), fog is much more common over inland areas. Flat coastal areas such as the airport at Campbell River are very prone to this type of fog. When the water temperature is lower than the air temperature (especially when lower than the dewpoint temperature), fog is much more common over the water.

Southeasterly winds tend to be strongest and most significant all along the coasts due to channeling and coastal convergence. Similar effects are observed with northwesterlies although these are less common.

Victoria To Vancouver/Nanaimo

(Map 1, opposite)

Summer and autumn weather is more conducive to flying; the main problem being sea stratus through Juan de Fuca Strait in the summer and sea fog in the fall. The San Juan Islands are more prone to morning fog than the Gulf Islands. The top of the fog usually lies between 1,000 and 2,000 feet AGL while the sea stratus tends to produce ceilings between 1,500 and 2,500 feet AGL. A typical pattern is for the stratus to move inland over succeeding days and become more and more reluctant to dissipate. Stratus typically lies just off Victoria Airport but covers the Gulf Islands. Fog banks off the east end of the runway at Victoria can be blown in as pulses by the seabreeze.

At the Duncan Airport, the runway is oriented 12-30 and lies uphill to the northwest. Westerly winds through Cowichan Valley often turn southward along the runway toward the valley and accelerate (15 kt wind becomes 25 kts). This downflow can be up to 500 ft/min resulting in aircraft landing short.

There are large hills on Pender Island where all winds except southerlies flow over hills resulting in downflow winds over the runway.

Vancouver Area Including Pitt Meadows, Langley and Boundary Bay

(Map 2, next page)

Weak surges of fog from Juan de Fuca Strait tend to cover half of Vancouver Airport and lie over Boundary Bay just off Boundary Bay Airport. Radiation fog is common around Langley and Pitt Meadows in the fall due to the low, flat bog land. Abbotsford Airport, however, is seldom a problem.

Boundary Bay area experiences the "San Juan effect" and will clear ahead of the rest of the Lower Mainland (likely due to subsidence behind the Olympic Mountains). When fog is advecting inland, Boundary Bay can get socked in very quickly.

IFR pilots report some of the worst icing conditions ever experienced occur over

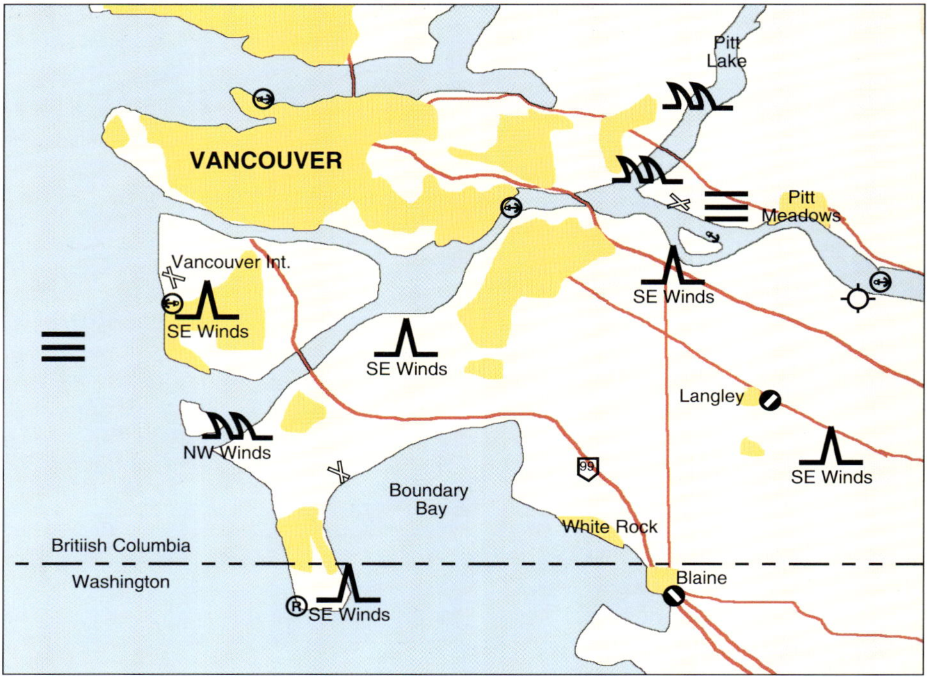

Map 2 — Vancouver area including Pitt Meadows, Langley and Boundary Bay

Vancouver's North Shore Mountains. Thus, most choose to ascend to the west over the Strait. Strong winds at Vancouver are usually from either the west or southeast with the southeasterlies tending to be gusty and turbulent. Westerlies can reach speeds exceeding 40 knots on occasion but tend to be stable and relatively smooth.

Tidal currents in the Fraser River can be hazardous for landing seaplanes, especially with northwest winds that oppose the current, producing steep, choppy waves. Similar or worse conditions are observed in the Pitt River, all the way to Pitt Lake.

Langley to Hope

(Map 3, opposite)

The worst weather in the Fraser Valley occurs from September to April and is usually associated with precipitation which brings on rapidly lowering ceilings and visibility. The low cloud tends to thicken noticeably near Agassiz and worsens as you go eastward into the mountains. In a southwesterly flow, poor weather piles up along the mountains with higher ceilings on the south side of the valley

During times of strong northwest winds over the Strait of Georgia, the winds will curl into a strong westerly along the valley. Westerlies of 25 to 45 knots with stronger gusts are common behind cold fronts. If the airmass is cold, snowshowers and very poor visibility will be found along the high ground that rims the south part of the valley. During times of strong inflow, anticipate turbulence on the south side of the Fraser River near Hunter Creek located west of Hope between Laidlaw and Flood. During times of strong outflow the weather will generally be clear but significant turbulence should be anticipated around Chilliwack and Sumas Mountains below 5,000 feet ASL.

The summer sea breeze channels between Chilliwack and Hope and can be quite strong (20 to 35 knots). Smog (top of layer 1,500 to 2,000 feet) can reduce visibility to 2 to 3 miles at the east end of the valley during warm, summer afternoons. When approaching from higher levels, reflection of sunlight off the smog can make it difficult to view the ground.

Very strong winds are common over Harrison Lake and Harrison River in summer. Over the lake, the worst winds are northerlies during summer afternoons. These can produce significant turbulence and waves as high as 4 feet.

Vicinity of Hope

(Map 3, opposite)

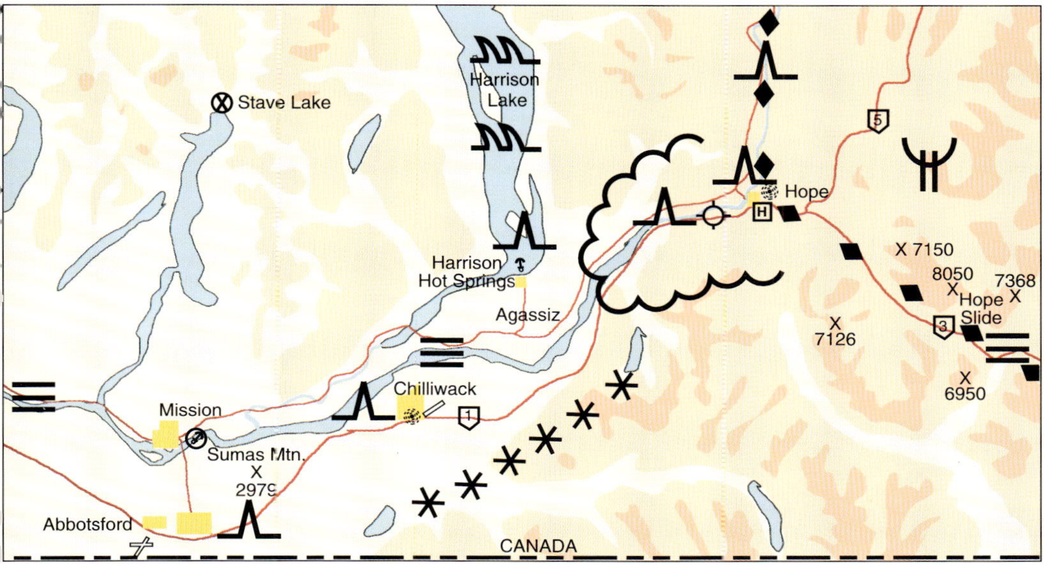

Map 3 — Langley to Hope

Low cloud and poor visibility are facts of life in Hope from late fall to late spring. Only during times of dry, cold outflow will the cloud clear from the area. Turbulence is common in the Hope area when strong inflow or outflow winds are occurring. Inflow is common year round while outflow tends be a winter phenomenon. Local pilots say that if you are above 3,000 feet then you will be out of the turbulence. During the winter months, icing in cloud is often significant to the east and northeast of Hope, along the Coquihalla to Merritt route.

O. Lange

Typical convective development along the west coast

(Maps 4 and 5)

Map 4 — Vancouver to Pemberton along Howe Sound

There is a high volume of traffic along this route throughout the year. Pilots can contact others on VHF to determine the weather ahead. There are some very high wires along Highway 99 which are dangerous for fixed wing aircraft following the river. Experienced pilots also tell of frequent navigation errors near Squamish, especially in bad weather.

In the summer, thunderstorms are not infrequent in Howe Sound and tend to move up the sound. Strong inflow winds are common into Howe Sound to Squamish, often exceed-

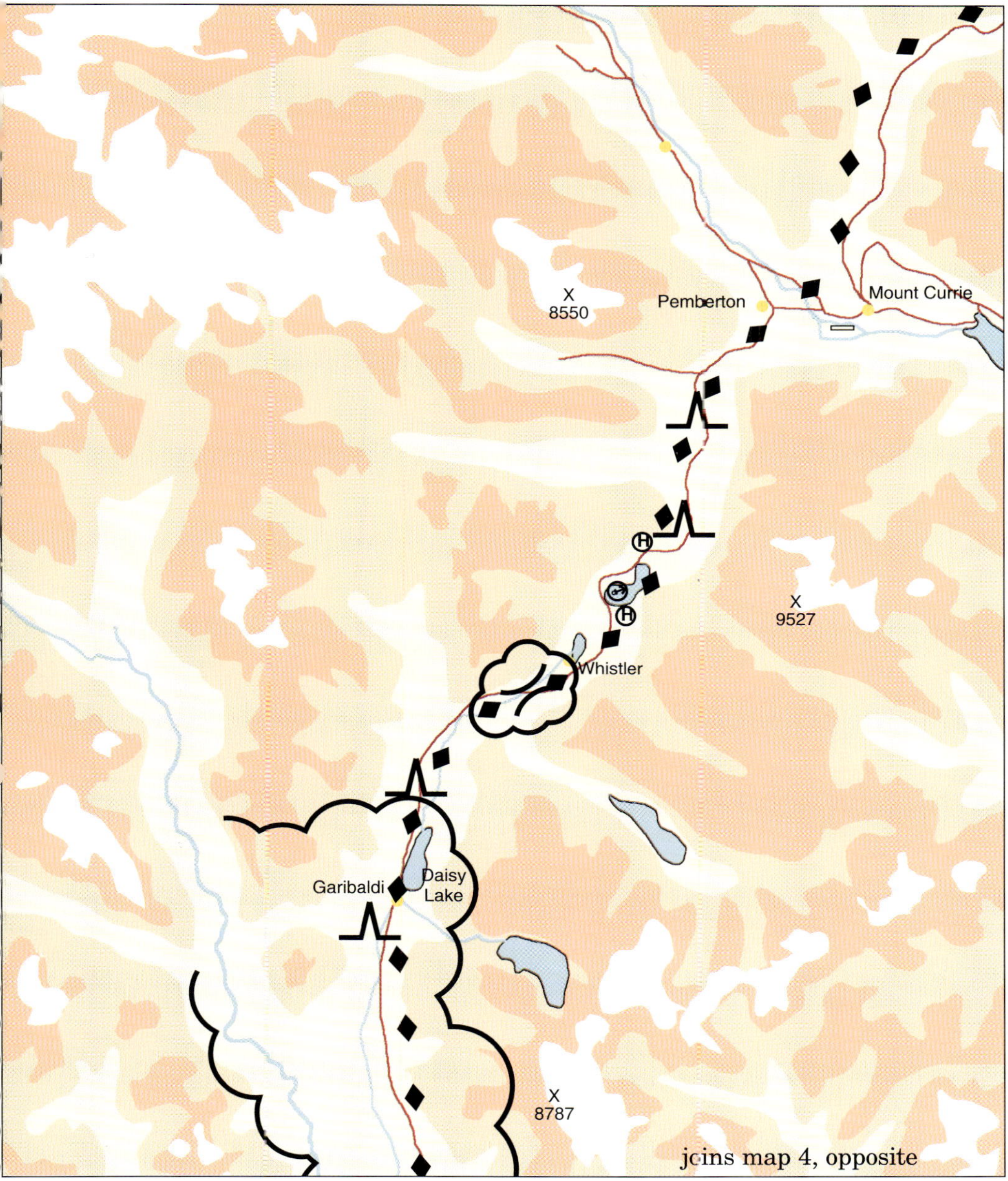

Map 5— Vancouver to Pemberton along Howe Sound

ing 30 knots.

In the winter, during outflow conditions, turbulence to 5,000 feet is common. Local pilots say the VFR route on the east side of the river is in the worst place. They recommend staying on the left side with the river to your right. Strong outflow winds blow through Queen Charlotte Channel while the area around Keats Island and Collingwood Channel can be calm. The worst location for low cloud is one to two miles north of Squamish. The low cloud usually ends near Daisy Lake. A second choke point exists near the bend in the road to the south of Whistler. In a strong southeast flow up Howe Sound, there is strong turbulence along the highway where it is cut into the cliffs.

(Map 6)

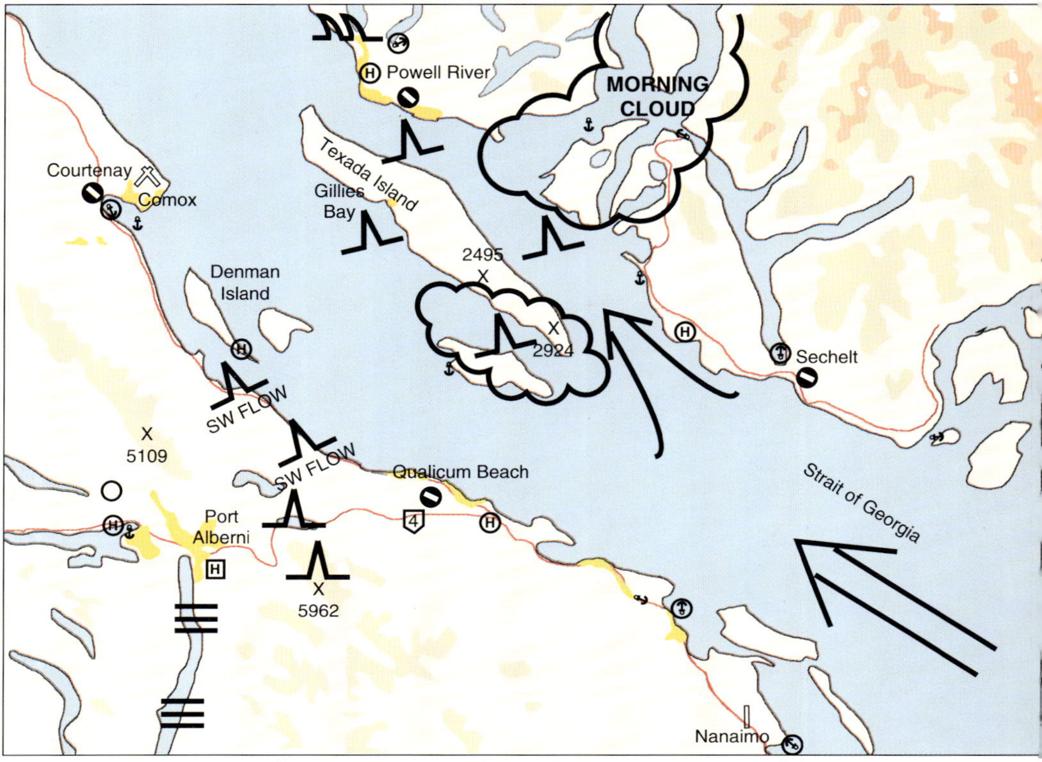

Map 6 — Straits Vancouver / Nanaimo to Powell River / Comox

Southeasterly winds are the most significant along the strait and they tend to be gusty. Turbulence along this portion of the strait is generally not significant except along the coast and the northern Gulf Islands. Channeling and coastal convergence can be severe along both sides of Texada Island due to the height of the ridge along the island. Coastal convergence can produce winds of up to 35 knots in the summer near Qualicum Beach. Nanaimo Harbour is almost always calm.

Low cloud and fog tend to persist wherever there are hooks of land to hold it, such as around Nanaimo and between Texada and Lasqueti Islands. In general, fog and low cloud are more persistent over the northern half of the strait. At Nanaimo, fog is much more common over the airport than at the floatplane base.

Powell River can be hazardous in southeast winds with lots of subsiding air near the runway. The lake can be especially bad in southeast winds which swirl around a point of land on the southern end of the lake. Although still dangerous at times, it is often calmer on the north end of the lake, and along the ridge on the west side of the lake.

(Map 7)

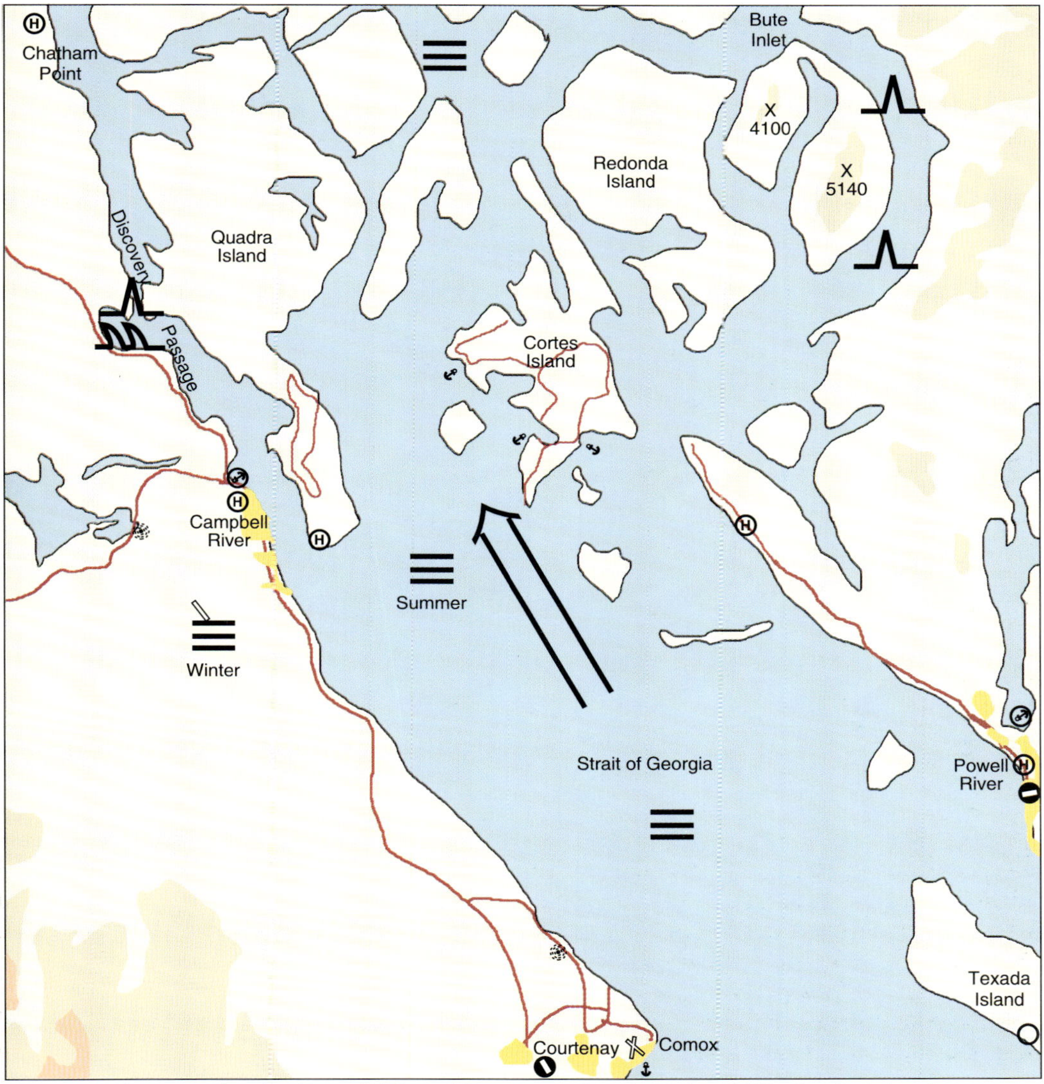

Map 7 — Straits Powell River / Comox to Quadra Island / Redonda Islands

As with areas further south, weather in this region tends to be relatively dry as Pacific weather systems are dried as they pass over Vancouver Island. The airport at Campbell River reports considerably more fog than occurs over the Strait, especially during autumn months. The Comox Airport, right along the water, is not affected as much by this fog.

Bute Inlet is very susceptible to strong outflow with vertical significant components over the water. Evidence can sometimes be seen in a "catspaw" area with whitecaps. A very turbulent spot, both on the water and in the air, is Seymour Narrows in Discovery Passage.

(Maps 8 west & 8 east)

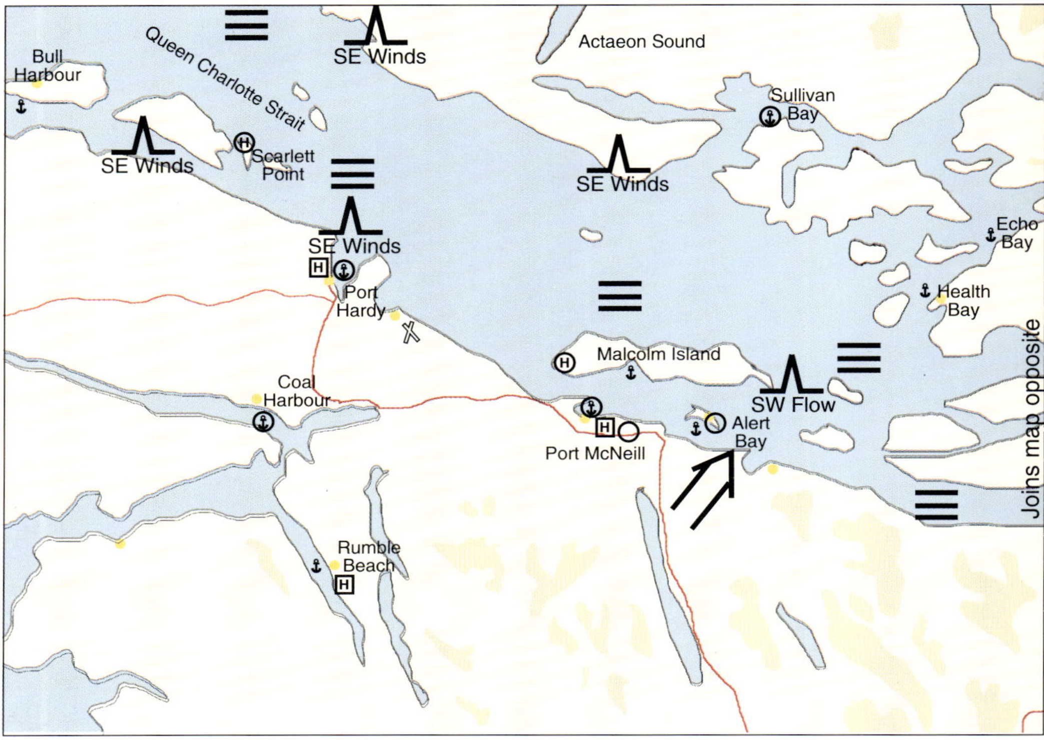

Map 8 west — Straits Quadra Island / Redonda Islands to Queen Charlotte Sound

In summer, other than the occasional weak frontal system, the main problem is fog which tends to sit along the coastal areas and over the straits. This situation is very common when there is an area of high pressure building to the west of the island. The fog and stratus become widespread overnight and generally wrap around the north end of the island and extend both northwards along the central coast and southwards to Malcolm Island or even as far as Chatham Point. Heating from the sun will dissipate the fog and stratus off the land by late morning however fog banks will persist over Queen Charlotte Strait throughout the day and begin to expand once the sun sets. In general during July and August, conditions tend to be poor in the morning, good in the afternoon and often decrease rapidly after sunset.

The sea breeze activity is pronounced in this area during the summer. Winds over the straits and at Port Hardy will be light in the morning and strengthen to northwesterly 15 to 25 knots in the afternoon. Alert Bay airport has an abrupt dropoff near the northwest end of the runway which induces a downflow in strong southeasterly winds.

Winter storms tend to induce strong southeasterly winds in the straits reaching typical values of 25 to 45 knots in Johnstone Strait and 35 to 50 knots in Queen Charlotte Strait. In such cases moderate to severe mechanical turbulence is common. Local pilots report that, in a strong ESE flow, severe turbulence is common near Actaeon Sound (about 20 miles northeast of Port Hardy), in Sargeant's Pass near Minstrel Island and in Surge Narrows (near Campbell River).

When the upper flow is strong southwesterly (around 50 knots or greater) there is a

Recommendation:
Aircraft trying to reach Port Hardy should ensure they have sufficient fuel to continue on to or return to Campbell River.

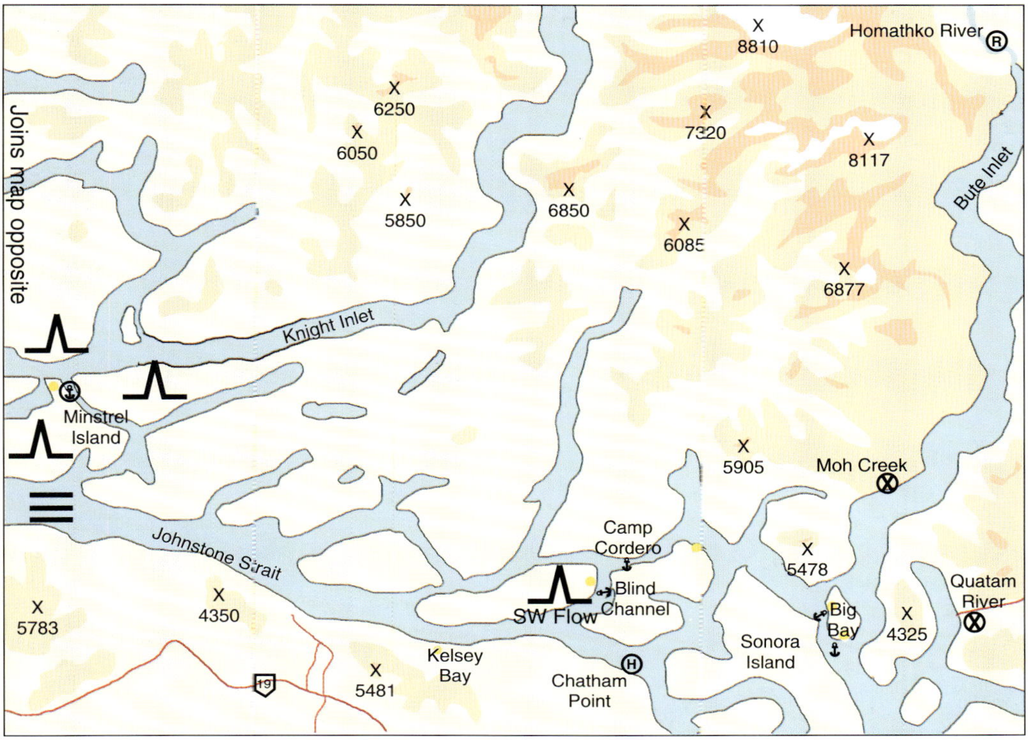

Map 8 east— Straits Quadra Island / Redonda Islands to Queen Charlotte Sound

tendency for strong channeled winds to come out of Telegraph Bay (near Malcolm Island). Also in such conditions strong rotor activity develops in Blind Channel (between Kelsey Bay and Chatham Point) resulting in swirling winds, sheets of water lifting from the surface and even waterspouts.

There is a typical weather regime change at Kelsey Bay, cloud and rain becoming more common further north as wind patterns change. One area with consistently poor weather is along Johnstone and Queen Charlotte Strait with the area between Chatham and Alert Bay receiving special mention for low hanging cloud.

The ceiling report at Port Hardy is often representative of surrounding land areas, as long as the differing terrain height is considered. If the ceiling is 500 feet at Port Hardy, terrain to the west is most likely obscured. In autumn and winter months, the northern end of the island is often battered by extreme winds reaching 80 knots or more. In such a case, turbulence is often extremely violent in the vicinity of the Brooks Peninsula and near Sartine Island in the Scott Islands Group.

West Vancouver Island Including Routes to Port Alberni and Tofino

(Map 9)

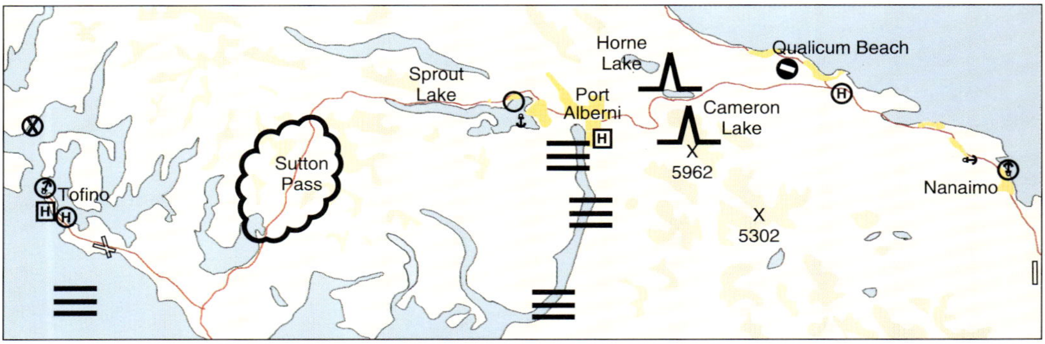

Map 9 — West Vancouver Island including routes to Port Alberni and Tofino

Note: Many references in this section are not shown on the map.

Some of the worst weather in British Columbia occurs along the west coast and western slopes of Vancouver Island. Strong winds, low cloud, heavy precipitation and fog banks combine to make this a treacherous route year round. Tofino is a favourite place to visit but low cloud and fog frequently sit off the coast and are quick to move over the airport. Local pilots say it is essential to have sufficient fuel to return to an alternate when trying to fly into the area. The observing site at Tofino Airport does not provide an unobstructed view to the sea and thus sea fog may not be seen. Lighthouse reports at Amphitrite Point, Estevan Point and Lennard Island are extremely important, especially for float plane activity. If any low cloud is reported at coastal airports, the west coast of the island is likely closed to VFR traffic.

Near Gold River and Tahsis, inflow winds can make water very rough at the seaplane dock. There are also strong river currents near the dock at Gold River.

The route from Parksville to Port Alberni may appear clear until just prior to reaching Port Alberni. Mechanical turbulence is common at the junction of Cowichan Lake Valley and Nitinat Lake Valley as well as shears occurring at the east end Cowichan Lake during afternoon when westerly flow along valley meets onshore sea breezes. The valley is tight from the southwest end of Cowichan Lake to Port Alberni with low cloud persisting sometimes at the bend in the road on the west side of the knoll. If flying from Alberni down Sprout Lake, light to moderate turbulence is common near Kennedy River. In a westerly flow, turbulence is frequent over the ridge near Cameron Lake as well as the area between Horne and Cameron Lakes. In the winter, conditions in Sutton Pass can get very low especially where the road sits up on the side of the pass. During the winter, snow along the road at the top of the pass is not uncommon.

Northern Vancouver Island to McInnes Island

(Map 10, opposite)

The scenery along the Central Coast is considered to be some of the most beautiful in British Columbia. Like the Port Hardy area, the weather in this region varies strongly between summer and winter, and the combination of widely scattered airports, lack of information and the speed at which the weather can change makes this a treacherous place to fly.

In summer, the main pattern is the persistence of low cloud and fog banks all along the coast from Prince Rupert to Port Hardy. These fog banks will move offshore with daytime heating but are capable of moving back in abruptly and without warning. Usually the fog

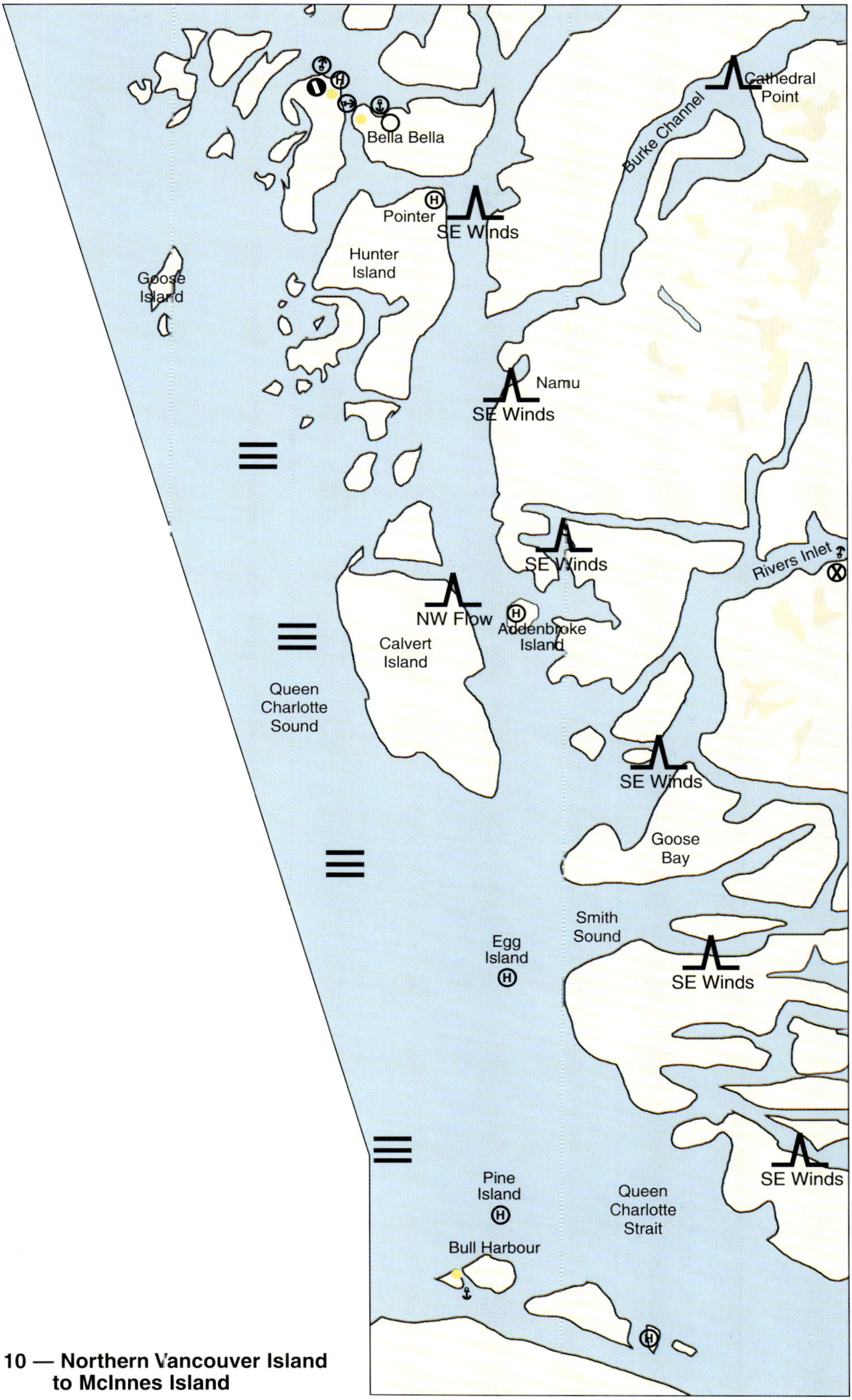

Map 10 — Northern Vancouver Island to McInnes Island

will only close the mouth of the inlets but in the extreme cases the entire inlet will fill up.

A strengthening afternoon inflow is common in the channels so that by late afternoon the winds are frequently strong and the water too rough to allow floatplanes to land at the head of inlets.

A couple of places have made a name for themselves. Local pilots suggest you stay at least 5 miles clear of the Goose Bay area (Namu) and the big hill on the south side of the mouth to Rivers Inlet when strong southeast winds are blowing. Lee waves are present over top of Rivers Inlet and flight at near 500 feet will be extremely rough in windy conditions.

Very strong outflow conditions are common when pressures build over the interior and values in excess of 50 knots are not uncommon near Cathedral Point in the Burke Channel. Outflow of cold air will also occur ahead of approaching systems and this creates the potential for freezing rain.

The weather report from Bella Coola cannot be used as representing Bella Bella. Bella Bella, the more exposed marine site, reports much more low cloud and fog *(see also map 40, page 101)*. Manned lighthouse observations provide vital weather reports for coastal flight planning. Westerly winds tend to produce severe turbulence over portions of Rivers Inlet due to the steep bluffs on the west side of the inlet. Sea fog is very common over Queen Charlotte Sound and in Queen Charlotte Strait. The lighthouse report from Egg Island is very representative for conditions along the mainland coast.

Bella Bella to Prince Rupert

(Map 11)

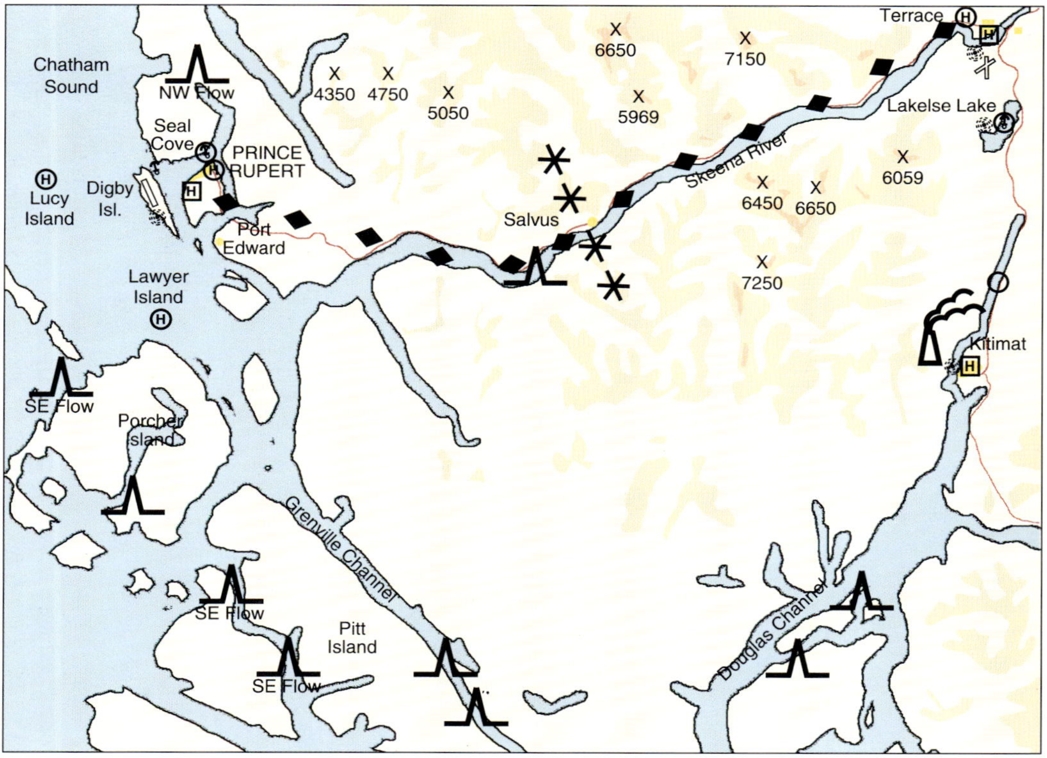

Map 11 — Bella Bella to Prince Rupert / Kitimat

As with most other regions of the coast, weather hazards differ from summer to winter. In winter, Pacific low pressure systems can produce very strong winds, often exceeding 50 knots.

When winds exceed 25 knots, moderate or greater turbulence can be expected around virtually all terrain. Channeling and funneling can produce extreme winds in most inlets. Even when winds reach 35 knots in Chatham Sound, Seal Cove can be relatively calm. Fog is more common over Port Edward and Digby Island than over Seal Cove.

In summer, two factors dominate local weather; sea fog and afternoon inflow seabreezes. The inflow can reach 40 knots in some inlets. Fog is most prevalent in August and early September. Throughout the summer the fog can sit over the sea, expanding right to the head of the inlets overnight and retreating during the day. Persistent fog often lies just west of Digby Island and can intermittently roll in and out over the runway or approaches.

Bella Bella to Kitimat (Douglas Channel)

(Partly shown on map 11, opposite)

When active Pacific storms are moving northward along the Central Coast, there may be rotor turbulence on the east side of the valley due to the higher southeasterly storm winds interacting with the generally north to south terrain. This turbulence is often severe if a strong southerly inflow is blowing. The worst turbulence is usually found from mid-mountain to peak levels within one or two miles of the slopes. Flight conditions will generally be smoother and safer along the west side of the passage.

Prince Rupert to Stewart

(Map 12)

Significant turbulence is observed with northwesterly winds to the north of Seal Cove. The channel at Kincolith experiences severe turbulence in inflow and outflow patterns. The Portland Canal to Stewart is also often very rough, especially in winter outflows. Above the highest terrain height, this mechanical turbulence is reduced.

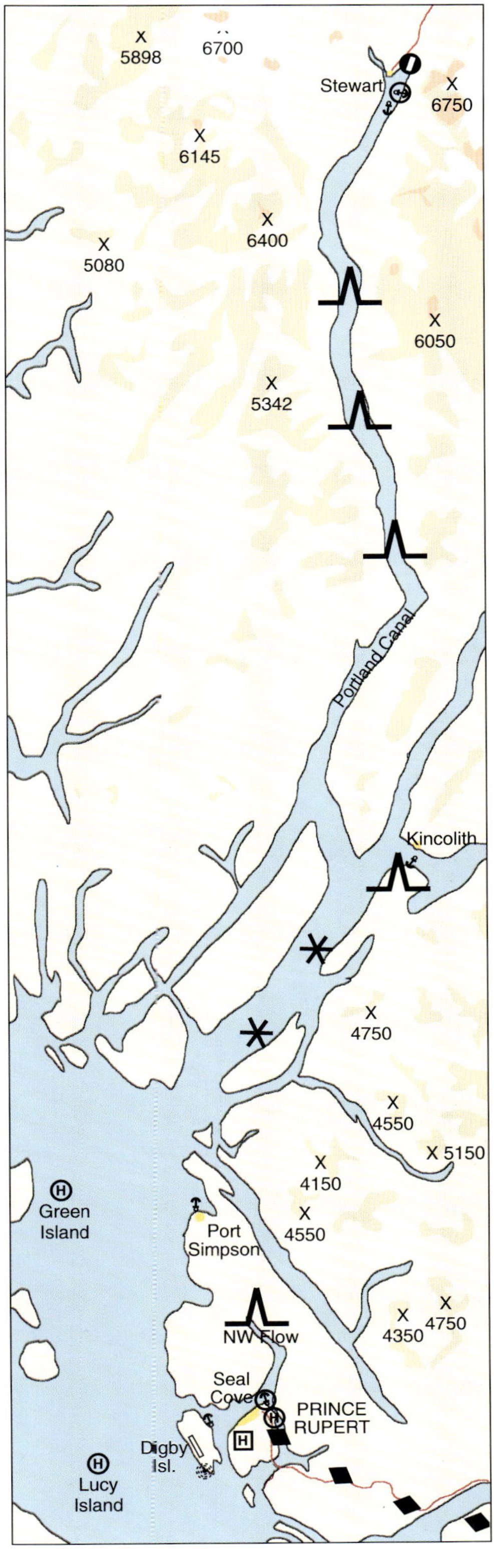

Map 12 — Prince Rupert to Stewart

Prince Rupert to Terrace/Kitimat

(Map 11, page 68)

It can be very rough behind all the shoulders along the river with Telegraph Point renowned for its severe turbulence. The turbulence can generally be reduced by staying over the river. The region near Salvus can receive extremely heavy snow. The airport at Terrace is more prone to low cloud due to its elevation, approximately 500 feet above town.

The mills and smelter at Kitimat can reduce visibilities and ceilings, particularly along the west side of the valley between Kitimat and Terrace.

Queen Charlotte Islands

(Not shown)

The Queen Charlotte Islands exhibit many unique weather phenomena which are beyond the scope of this manual. Experienced pilots are the best source of information. Moderate turbulence is often reported to the lee of Tow Hill, east of Masset, during northwesterly or southeasterly winds exceeding 25 knots. Thunderstorms (summer and winter) tend to move through Skidegate Channel on a regular basis reducing visibility from unlimited to near zero (rain or snow) in a matter of a few minutes.

O. Lange

Estevan Lightstation, western Vancouver Island

Dawson Creek to Watson Lake

Northeastern B.C. experiences weather which can vary between typical mountain weather to Prairie conditions, depending on the overall flow pattern. Lee wave and associated turbulence form as in southwestern Alberta. However, given that the mountains are slightly lower and the range considerably wider, these lee waves possess less energy than their southern counterparts. However, the ride will be very rough, up to approximately 8,000 feet, if winds across the mountains exceed 25 knots. The lee wave activity does not generally extend north of 59 degrees where the mountains lose their linear characteristic.

Another phenomenon which is typical to the Prairies is the development of a low level nocturnal jet. Most common in spring and summer months, this feature occurs on clear nights following strong, gusty winds in late afternoon. As the sun sets, a low level temperature inversion forms. The strong winds remain several hundred feet off the ground while surface winds become calm. In some cases, the winds just above the ground can be stronger than the afternoon gusts. The inversion deepens during the night to as high as 1,000 feet above ground, before being burned off the following morning.

Patchy freezing rain or drizzle occurs on occasion, in winter months, east of the mountains, especially near the Arctic front when it lies along the Continental Divide.

K. Johnson

Prairie thunderstorm at night

Dawson Creek to Fort Nelson

(Map 13, opposite)

From Dawson Creek to Fort St. John (see map 16 east, page 79) the route along the highway is relatively flat. However, low cloud may block the portion 10 to 15 miles south of Fort St. John as this is slightly higher ground. Morning fog, which often forms along the Peace River Valley tends to be more frequent at Fort St. John than surrounding airports. The fog usually extends to just north of the Fort St. John Airport.

The route further north along the Alaska Highway is much more prone to low cloud in low level flow from the northeast quadrant. The high ground between Wonowon and Prophet River may be obscured while Fort St. John and Fort Nelson both report clear conditions.

Hills to the east and mountains to the west form a natural lowland area along Beatton River and the railway to Fort Nelson. Many pilots choose this route, especially in upslope flow when the surface wind is from the north to east quadrant.

Both Dawson Creek and Fort St. John can experience Chinook conditions, similar to those in southern Alberta and northern Montana. These are even more frequent closer to the Rockies, at Chetwynd. With the presence of an arctic high pressure system to the east, an intense shear zone can develop as strong southwesterlies aloft blow over surface easterlies. The shear zone is usually accompanied by a very sharp temperature inversion. In some cases, surface temperatures can be twenty degrees colder than temperatures several hundred feet above ground. In such cases, even in clear air, frost can form, particularly on the cockpit windows, while ascending into the warmer air.

Weather Hint: The smoke from the larger mills often can demonstrate the low level wind and shear pattern.

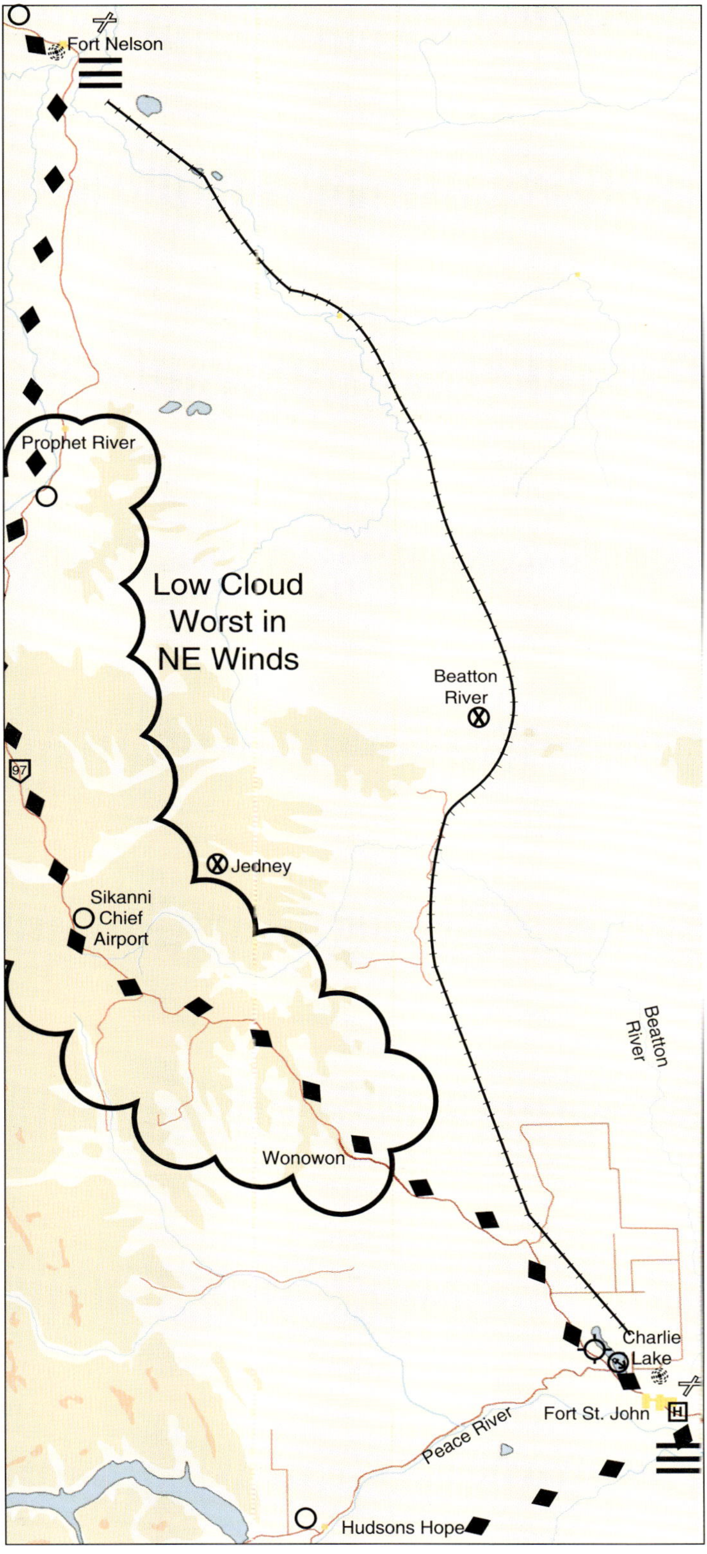

Map 13 — Dawson Creek to Fort Nelson

Fort Nelson to Watson Lake

(Maps 3, west and east

The Fort Nelson Airport is in a hollow and thus experiences more frequent fog than surrounding areas. It also has rather light winds compared to surrounding regions.

Some of the sharpest low level temperature inversions occur over Fort Nelson in winter, sometimes as much as a 30 degree change in less than one thousand feet. This can cause significant fog or frost on exterior surfaces on ascent.

The weather between Steamboat and Liard River often bears little resemblance to that at surrounding weather stations. If low cloud is present at Fort Nelson or Watson Lake, this route is almost certainly closed. Even with clear skies at both ends, the route may be impassable. The worst area is typically between Summit Lake and the pass west of Toad River. Most experienced pilots head northwest from Fort Nelson along the airway until reaching the Liard River then following the river to Watson Lake.

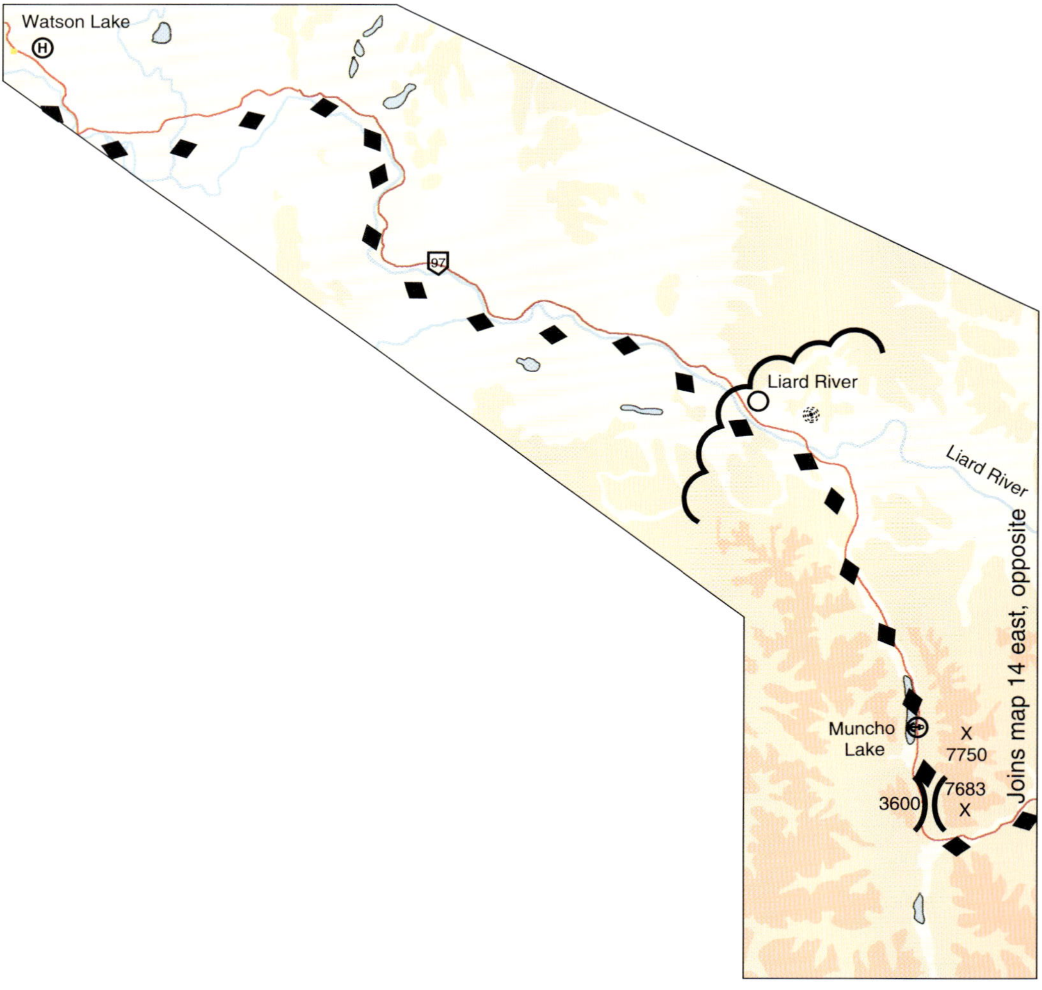

Map 14 west — Fort Nelson to Watson Lake

Cirrus cloud

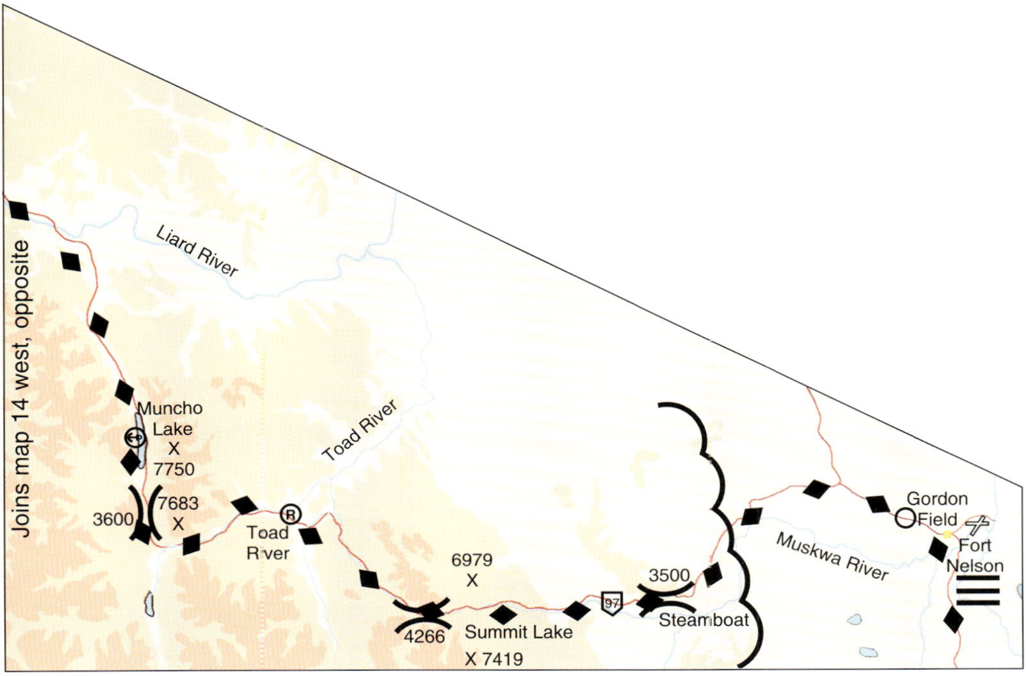

Map 14 east — Fort Nelson to Watson Lake

Crossing the Rockies Dawson Creek/Fort St. John to Mackenzie

(Maps 15, opposite, and 16 east and west, pages 78 and 79)

Turbulence in this region is worst in north to south oriented valleys when the winds are blowing across the mountains from west to east. These winds tend to be strongest in spring and autumn with associated mechanical turbulence usually below 1,000 feet AGL. The worst section along the indicated VFR route is near Pine Pass where low cloud, turbulence and very heavy snow are common. The Powder King Ski Hill near Pine Pass receives approximately 40 feet of snow each winter. This heavy snow occurs in a band from Pine Pass to McLeod Lake. In summer, the Pine Pass area is also prone to thunderstorm activity.

Thunderstorms often form along the foothills in summer months along a line just east of Chetwynd. These convective lines tend to remain relatively stationary while developing in the early afternoon then tend to begin to move in the direction of the upper level (approximately FL180) winds late in the day.

When traveling from the northeast to Tumbler Ridge, most pilots choose to take the route along the Murray River. From the southwest, a route from MacGregor River to Monkman Pass (Monkman Lake) then along the Murray River is utilized to get to Tumbler Ridge. There are, however, several dangerous canyons which can be mistaken for the main route.

The Murray River/Monkman Pass area generally experiences drier and calmer weather than Pine Pass, throughout the year.

Valley in the Central Interior of B.C. *G. Fricska*

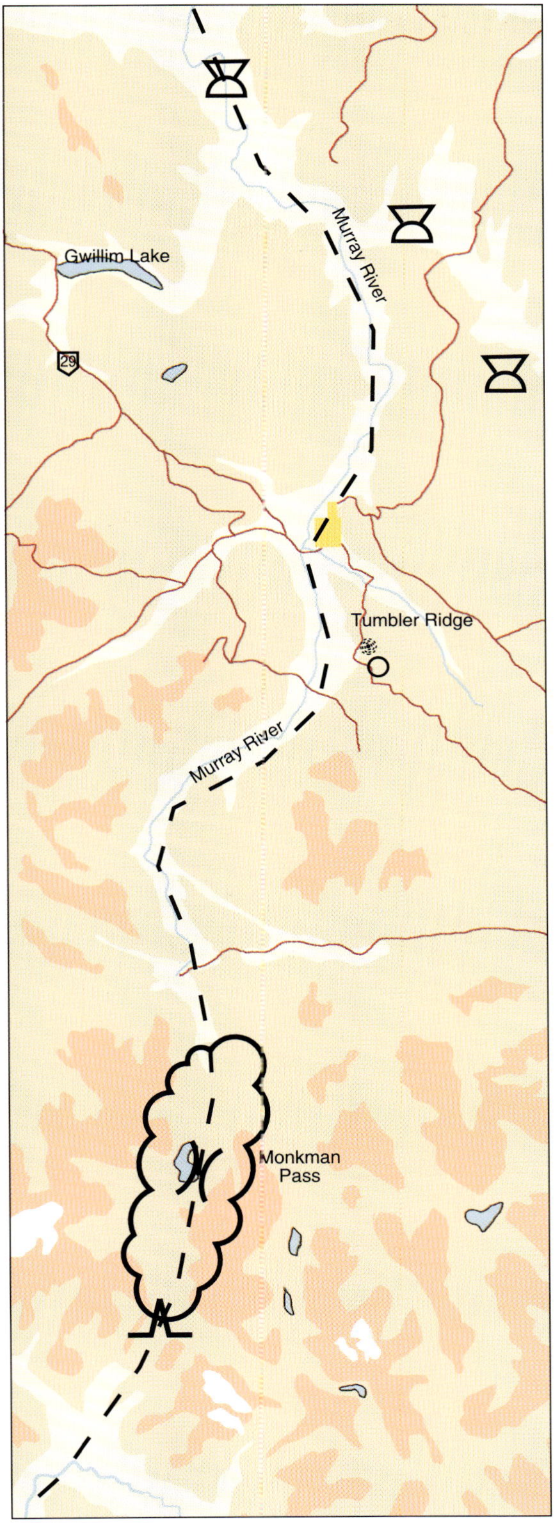

Map 15 — Crossing the Rockies Dawson Creek / Fort St. John to Mackenzie

Crossing the Rockies Dawson Creek/Fort St. John to Mackenzie

(Maps 16 west and east and map 15, previous page)

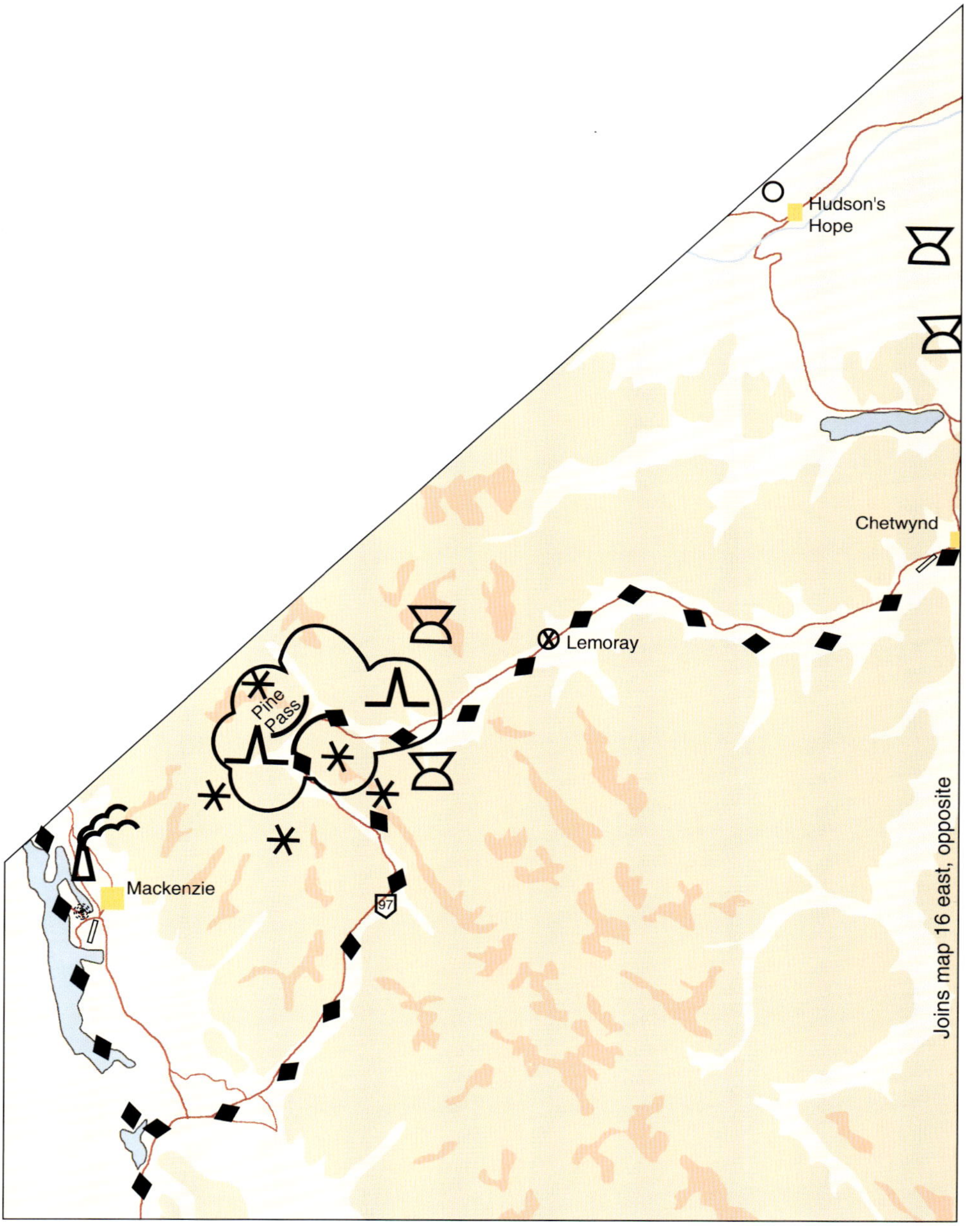

Map 16 west — Dawson Creek to Fort Nelson

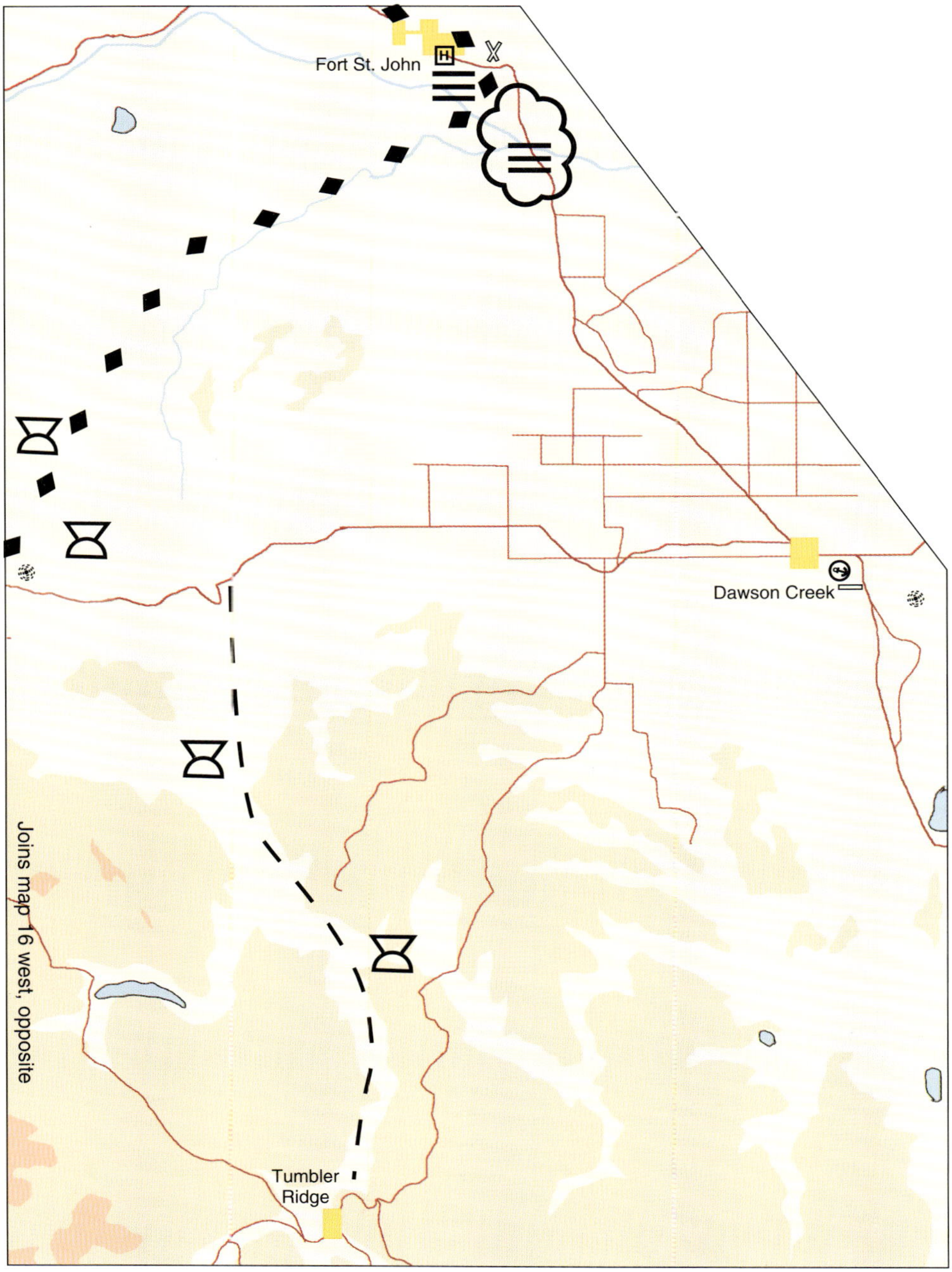

Map 16 east — Dawson Creek to Fort Nelson

Rocky Mountain Trench Watson Lake to Fort Ware

(Maps 17, north and south)

From Watson Lake southward, the terrain gradually rises with higher frequency of low cloud. Pilots must exercise caution not to mistake the Gataga River Valley for the Trench. Similarly, heading north, the Kechika River is often mistaken for the route to Watson Lake. The terrain continues to rise heading south to Sifton Pass, reputedly the worst spot along the Trench with frequent turbulence and cloud to the treetops.

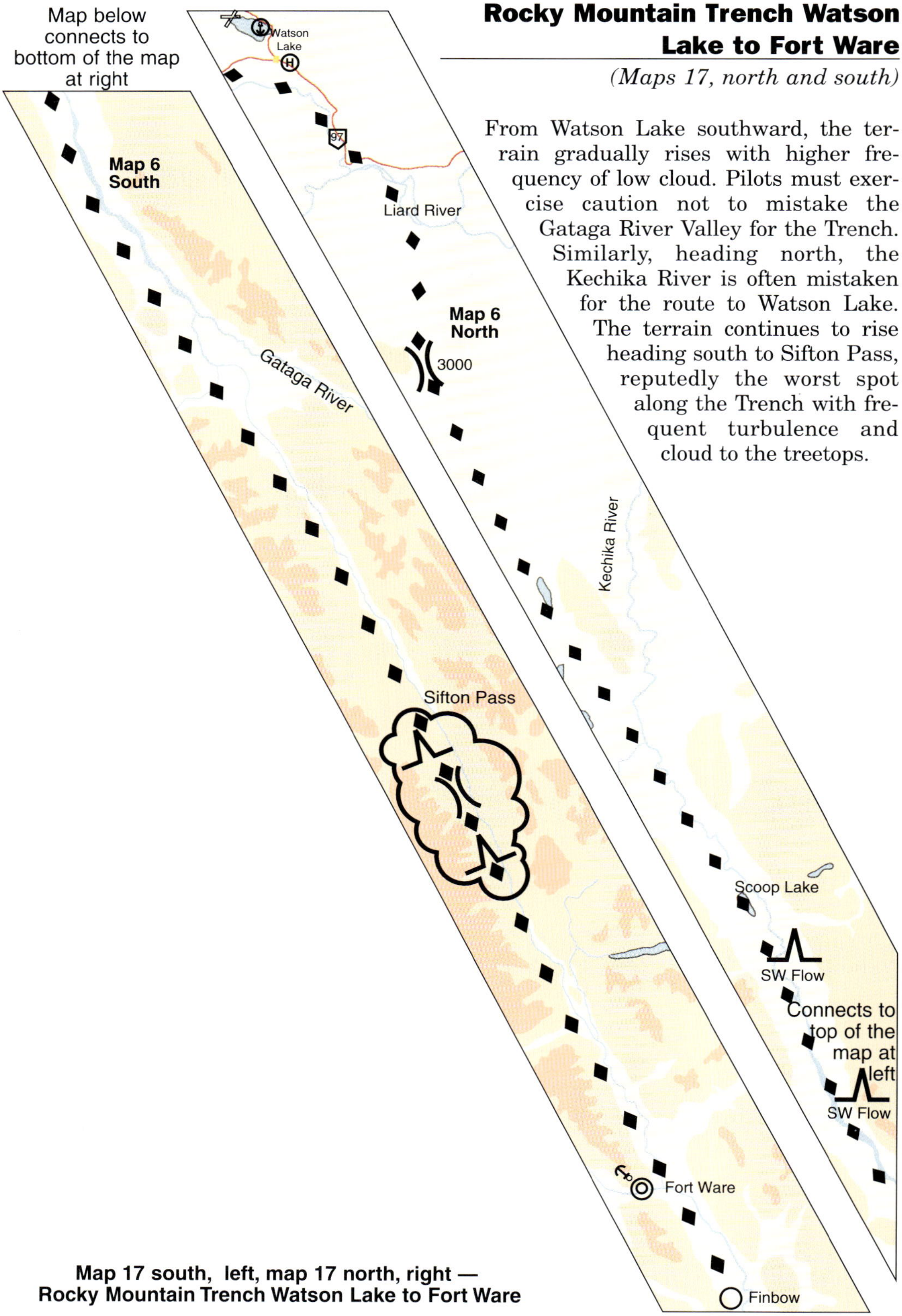

Map 17 south, left, map 17 north, right — Rocky Mountain Trench Watson Lake to Fort Ware

Rocky Mountain Trench Fort Ware to Mackenzie

(Map 18)

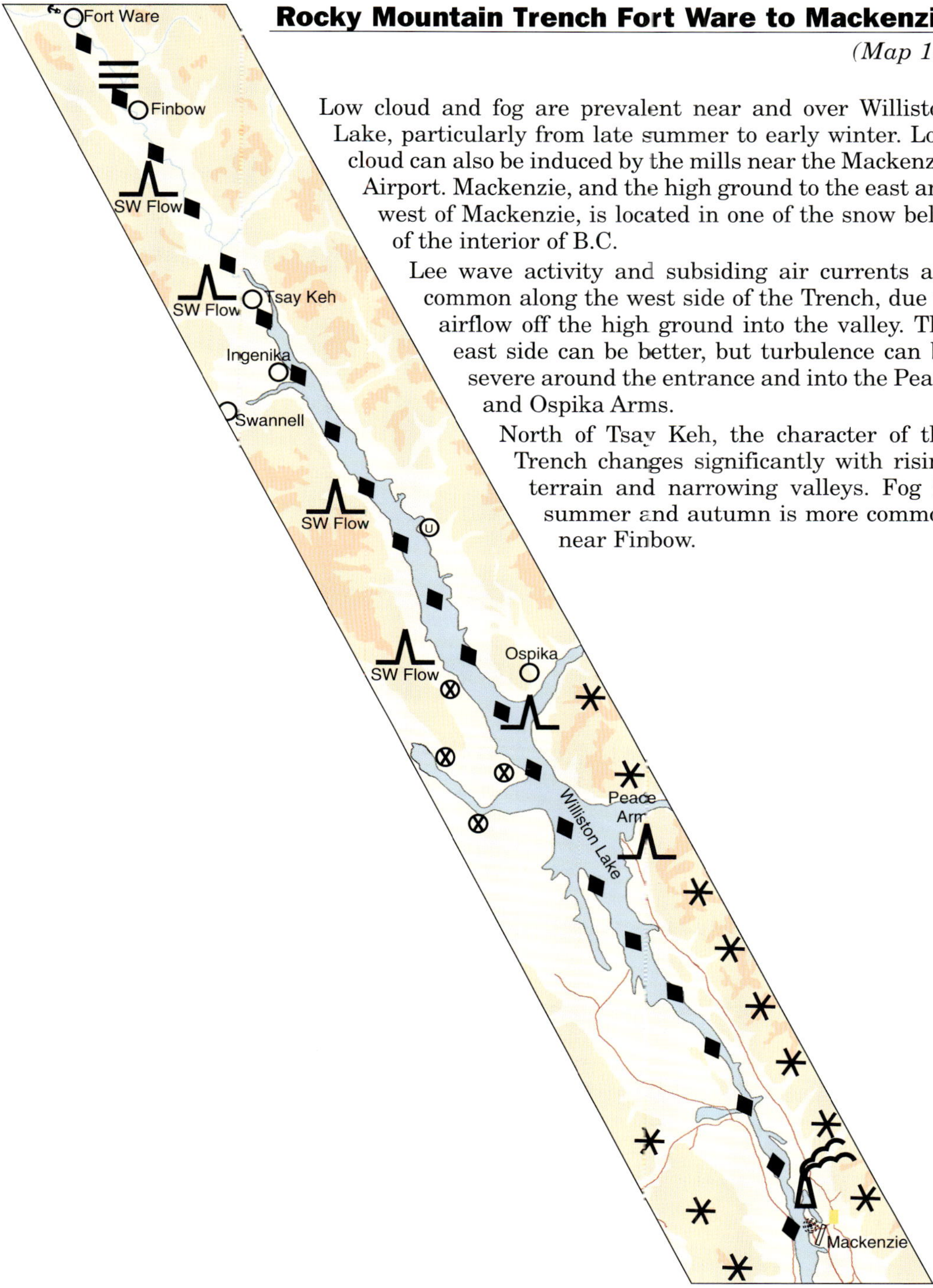

Low cloud and fog are prevalent near and over Williston Lake, particularly from late summer to early winter. Low cloud can also be induced by the mills near the Mackenzie Airport. Mackenzie, and the high ground to the east and west of Mackenzie, is located in one of the snow belts of the interior of B.C.

Lee wave activity and subsiding air currents are common along the west side of the Trench, due to airflow off the high ground into the valley. The east side can be better, but turbulence can be severe around the entrance and into the Peace and Ospika Arms.

North of Tsay Keh, the character of the Trench changes significantly with rising terrain and narrowing valleys. Fog in summer and autumn is more common near Finbow.

Map 18 — Rocky Mountain Trench Fort Ware to Mackenzie

Cariboo — Central Interior

For most of the year, aircraft operations are not hindered excessively by the weather. The notable exception is during the winter when the arctic front, accompanied by freezing drizzle, low cloud, fog and windshear moves back and forth through the area. The Cariboo-Central Interior is not prone to winds. However, the Fraser Valley can be windy, particularly southerly winds which can exceed 60 knots in low levels in autumn and winter. Although strong with high windshear values, these winds do not generally produce significant turbulence in the Valley.

Mackenzie to Prince George

(Map 19)

Low cloud and fog are often encountered near Summit Lake, extending as far as McLeod Lake at times. Winds and turbulence are generally light along this route. When going east up the McGregor River, look for strong turbulence near Monkman Pass *(see map 15, page 77)*. The onset can be very abrupt when approaching from the east.

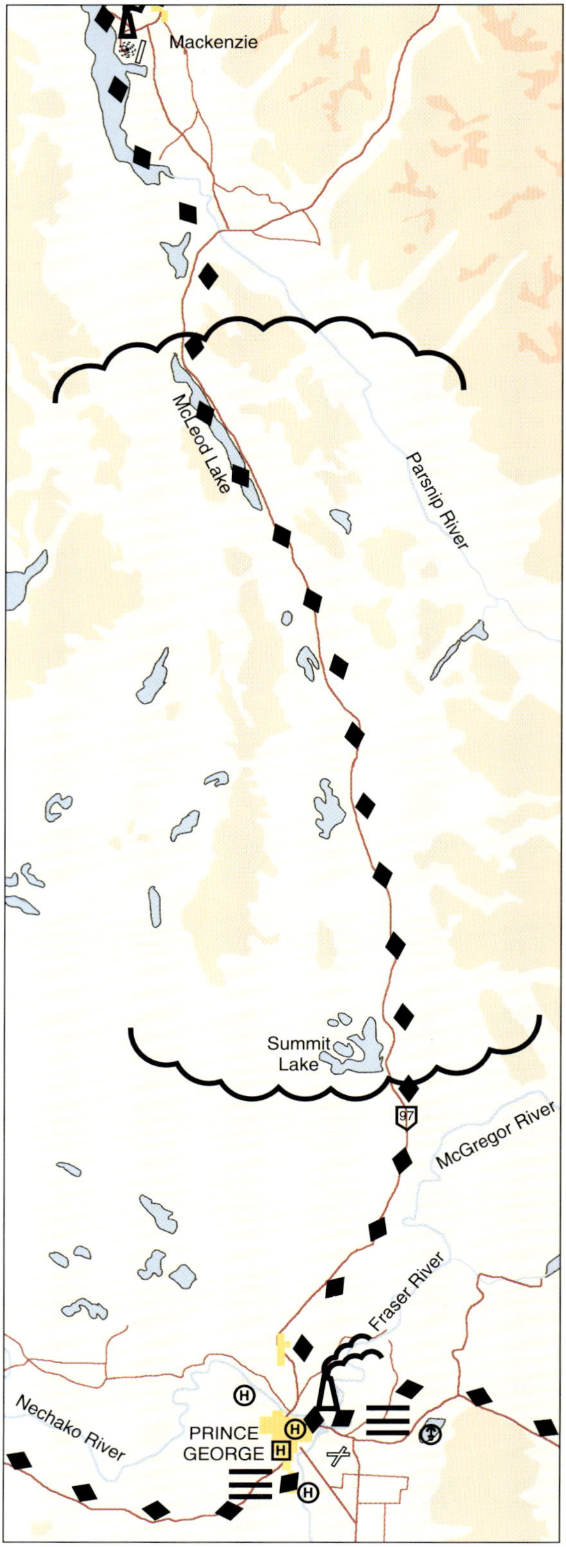

Map 19 — Mackenzie to Prince George

Prince George Area

(Map 20)

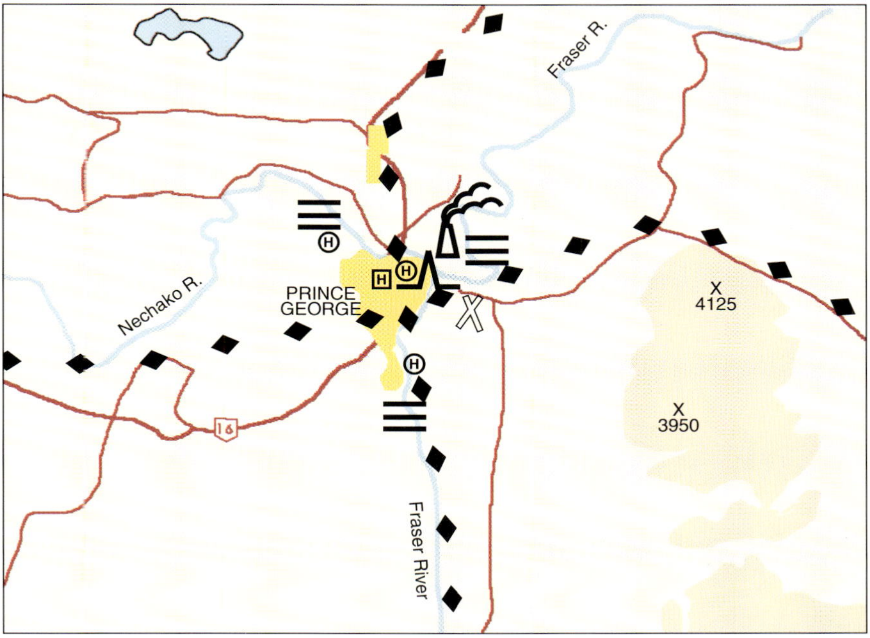

Map 20 — Prince George Area

The most significant hazard around the Prince George Airport is low cloud and fog, enhanced by the moisture along the rivers as well as the pulp mills north of the airport. Even on a clear day, fog associated with the mills can obscure the approach to runway 15. In this case, the southern approach to runway 33 is generally better. Strongest winds at the airport tend to be from the south. Light northerly winds at the surface often occur in conjunction with strong southerly winds aloft, say at 800 feet The loss of airspeed on take-off can cause small aircraft to drop into the river valley, near the bluffs on the north side of the river. This is most significant under morning inversions.

Prince George East to Fraser River at Goat River

(Map 21)

Strong low level turbulence due to funneled winds is observed along the Taspai Creek, just south of the main route.

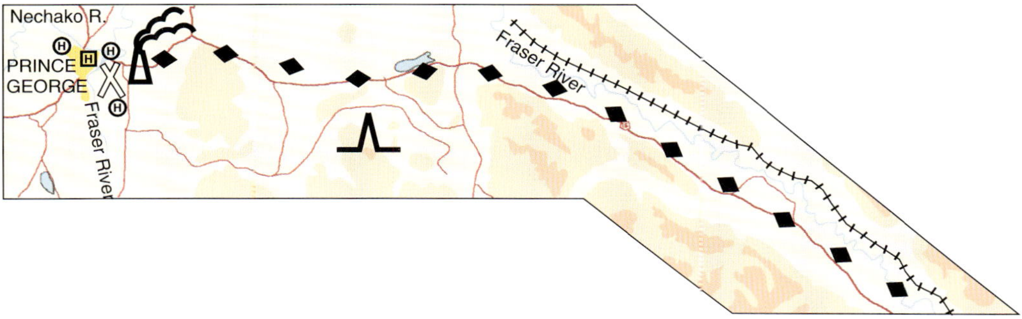

Map 21 — Prince George East to Fraser River at Goat River

Rocky Mountain Trench Goat River to Valemount

(Map 22)

Summer convective cloud is common along both sides of the Trench over the higher ground. Severe turbulence often results when strong winds aloft cross the mountains from the southwest. This turbulence does not generally extend above the terrain height, 6,000 to 8,000 feet. Severe turbulence is normal on summer afternoons over the glaciers to the northeast.

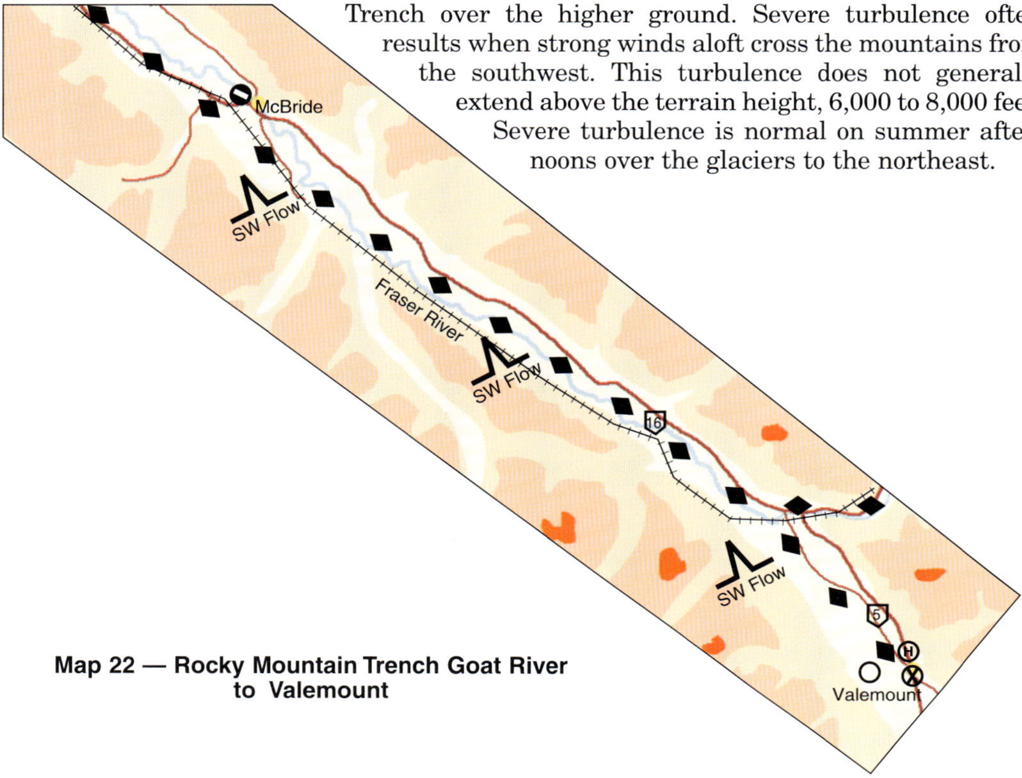

Map 22 — Rocky Mountain Trench Goat River to Valemount

Rocky Mountain Trench Tete Jaune to Mica Dam

(Map 23)

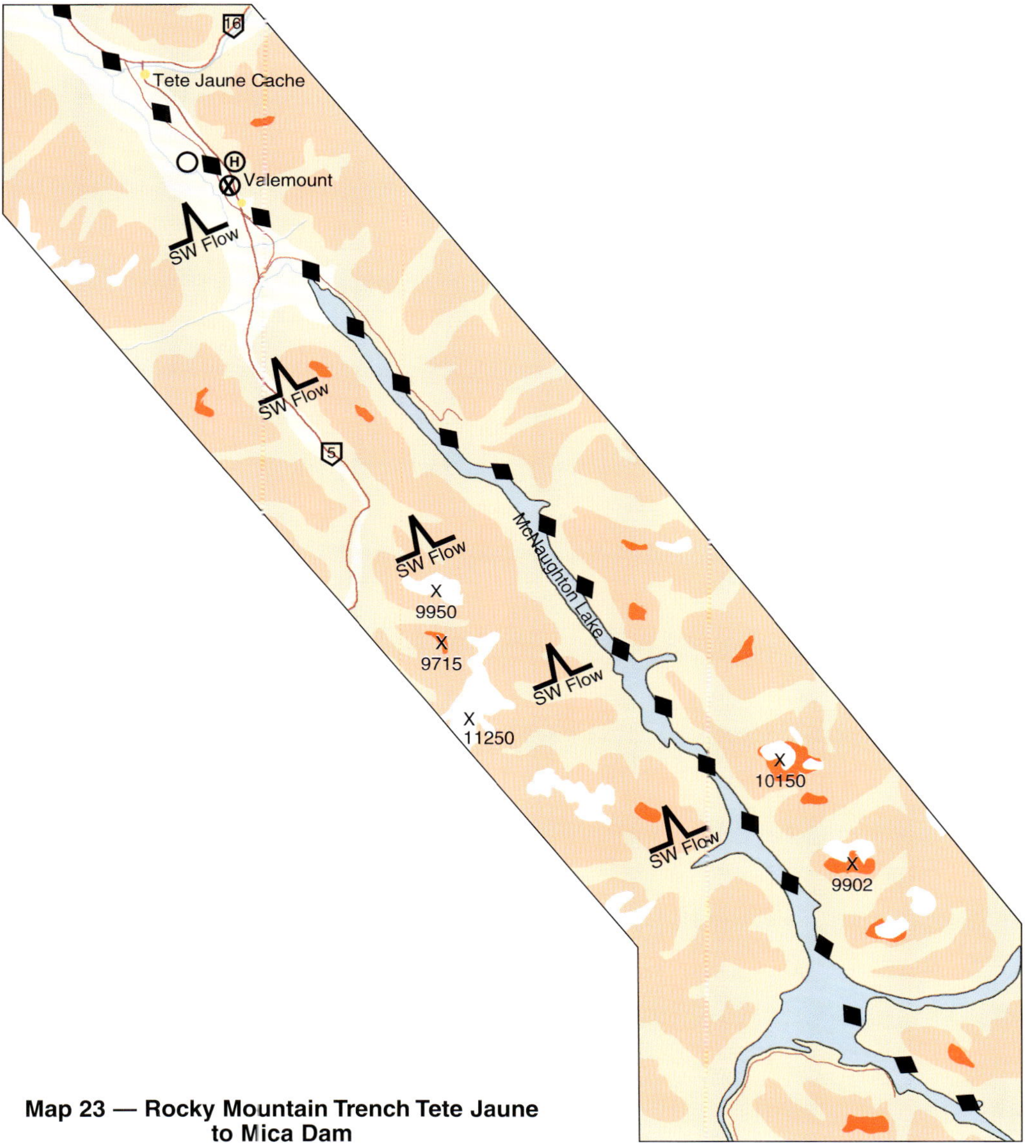

Map 23 — Rocky Mountain Trench Tete Jaune to Mica Dam

Rocky Mountain Trench Mica Dam to Donald

(Map 24)

Although still in the relatively dry Rocky Mountain Trench, this section receives considerably worse weather than areas further south. The main VFR route along Blackwater Creek is higher than an alternate route over the water; however, in bad weather, neither route is recommended by experienced pilots. Winds along the Columbia are almost always much stronger than those at Golden. The water usually freezes by mid winter, reducing the amount of fog and low cloud.

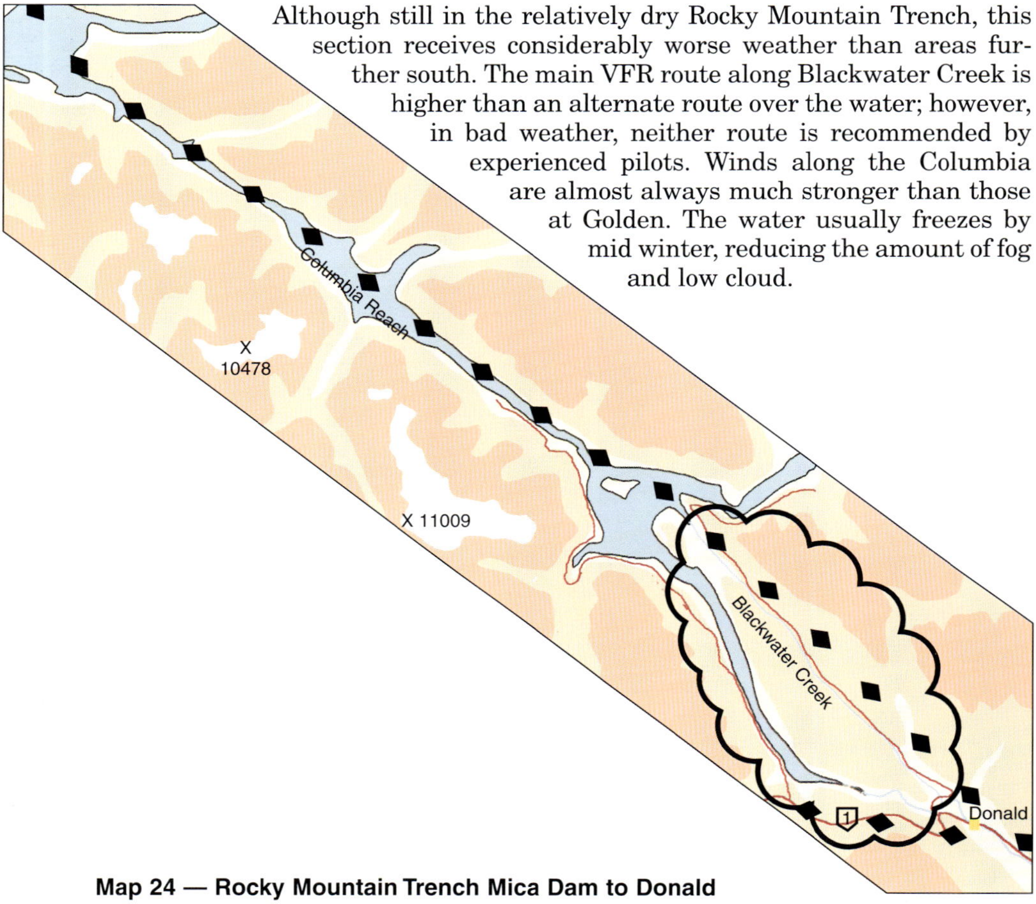

Map 24 — Rocky Mountain Trench Mica Dam to Donald

(Map 25)

This section of the Rocky Mountain Trench route may receive the best aviation weather in all of Western Canada. Weather-related hazards are few except for relatively frequent low cloud and fog near Parson. Infrequent strong winds do not usually produce significant turbulence due to the wide, linear valley, except where side valleys merge with the Trench.

Fog is common under clear skies in autumn although it generally burns off by 10 a.m.

Winter valley cloud between 4,000 and 6,000 feet often closes higher routes and produces significant icing.

> *Weather hint:* Golden Airport is not usually representative of the weather along the Trench. It tends to be much drier than areas to the north or south, especially north.

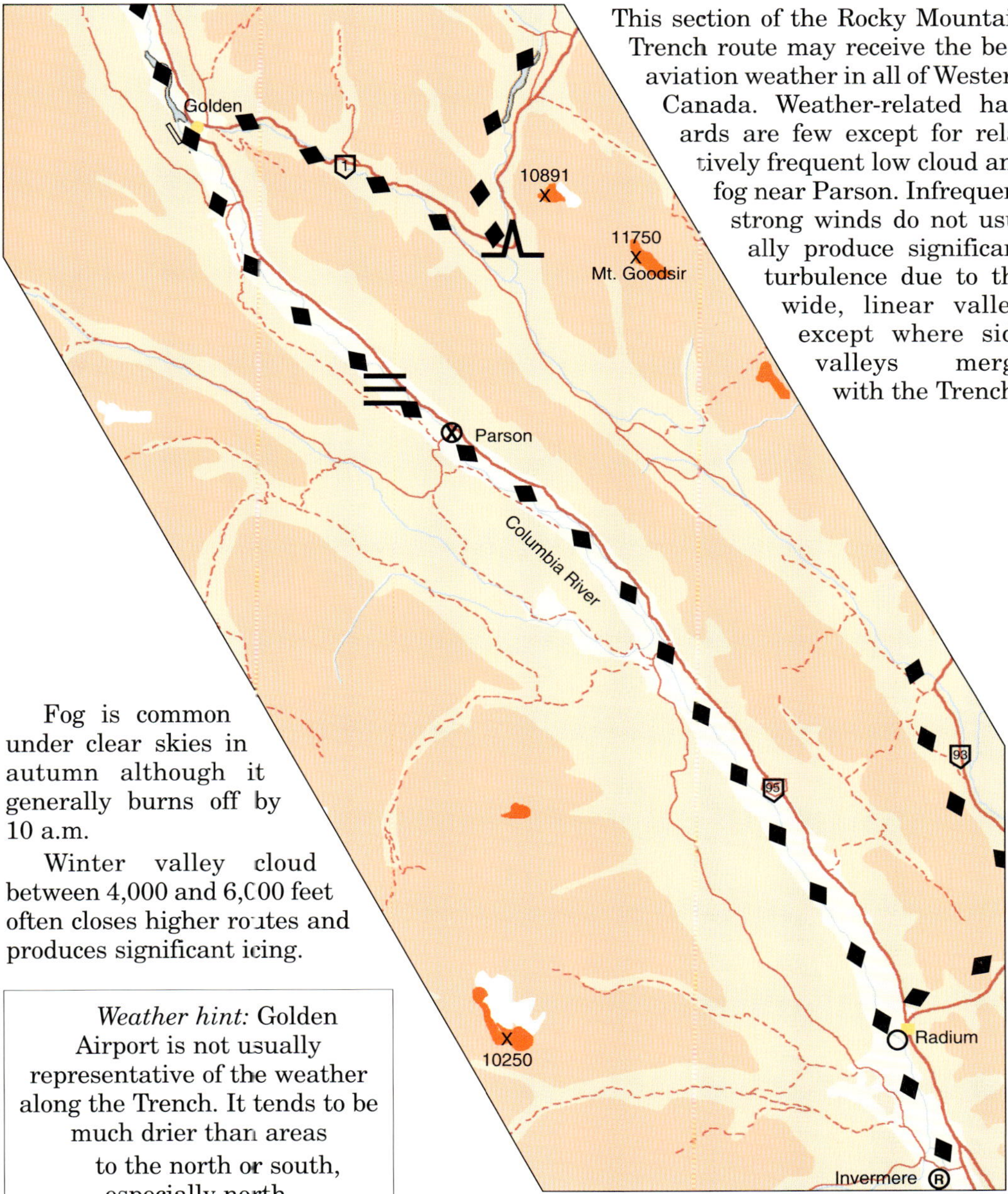

Map 25 — Rocky Mountain Trench Donald to Invermere

Rocky Mountain Trench Invermere to Cranbrook

(Map 26)

Conditions along this portion tend to be better than those further north.

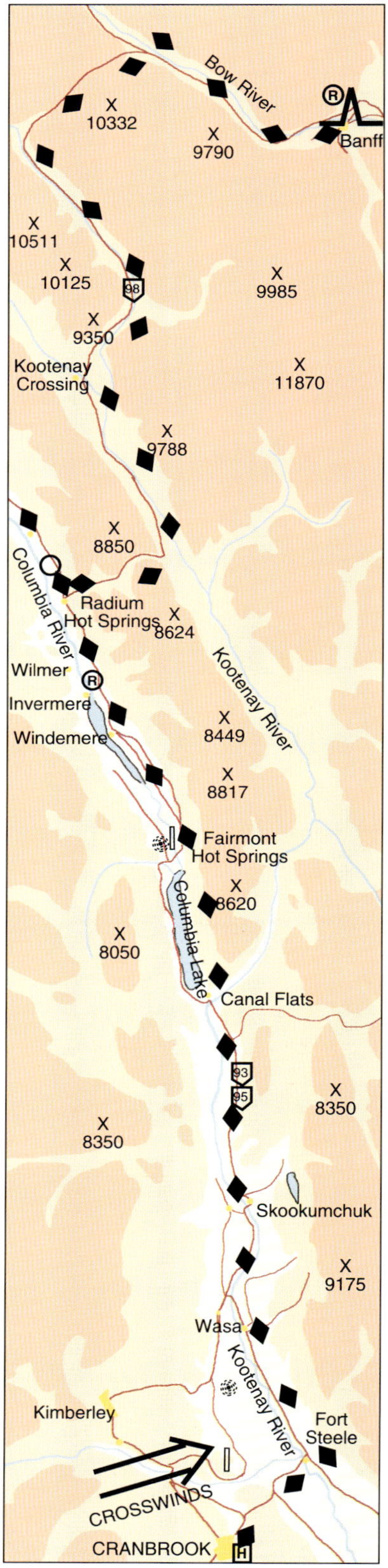

Map 26 —
Banff - Radium - Cranbrook

Tete Jaune to Jasper

(Map 27)

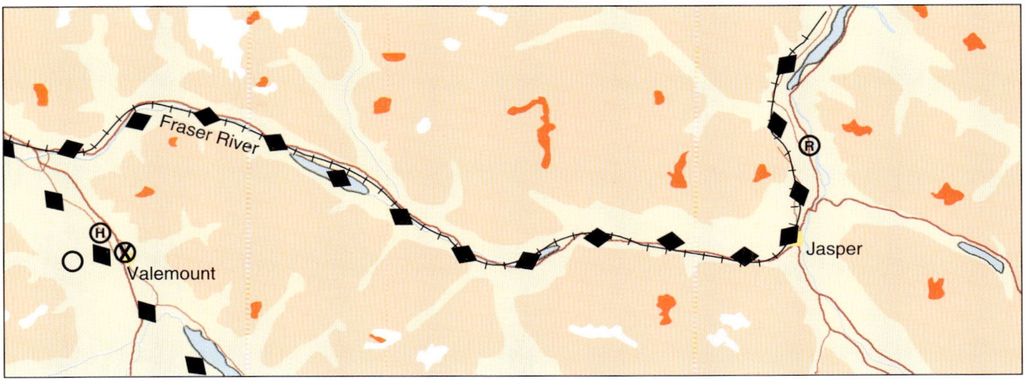

Map 27 — Tete Jaune to Jasper

K. Johnson

Near Jasper

Prince George to Quesnel

(Map 28)

Fog is relatively common along the valley, especially near Prince George and Quesnel, due to the mills located just north of the Prince George Airport and just south of the Quesnel airport. Some days, the fog will extend only a few miles from the airports with clear conditions elsewhere. The emissions from the mill stacks often give a visual representation of the low level vertical wind pattern.

Turbulence is not usually strong along this route expect near Trapping Lake up to 8,000 or 9,000 feet. In summer, convective cloud is most common in a southwest to northeast band through Hixon.

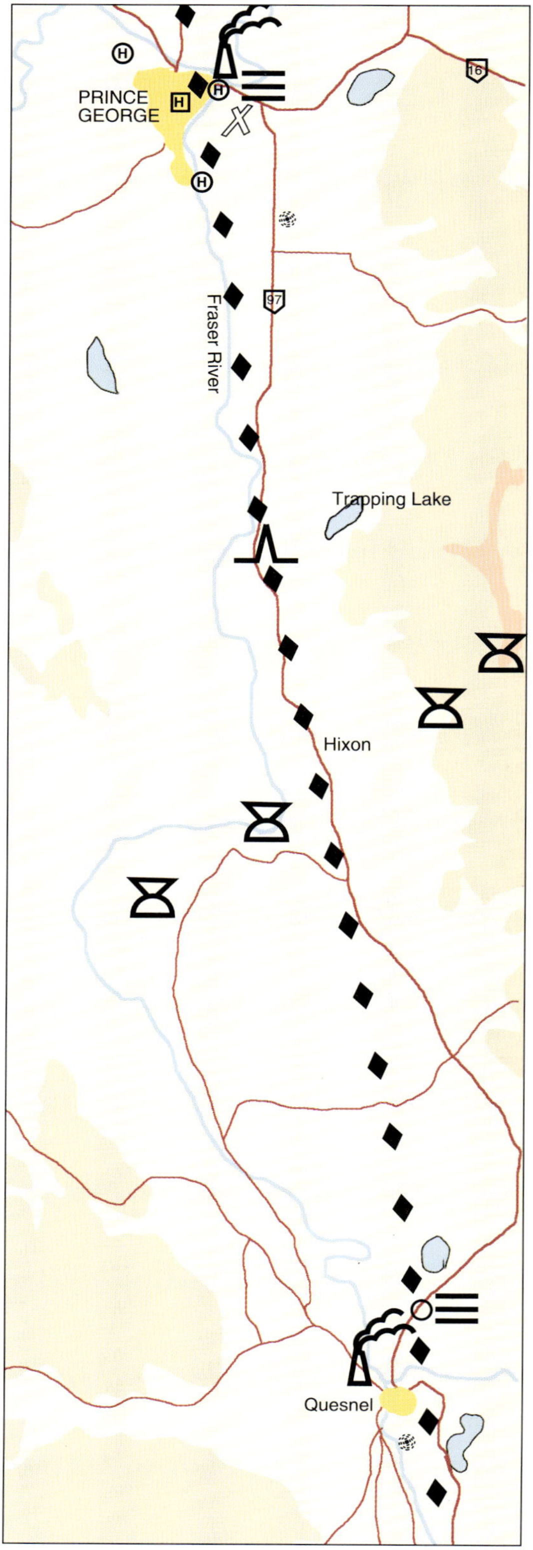

Map 28 — Prince George to Quesnel

Quesnel to Williams Lake

(Map 29)

Fog that fills the valley around Prince George and Quesnel is often below the much higher altitude Williams Lake Airport. However, it must be recognized that, in cases of extensive low cloud, the valley may be suitable for VFR flying while the Williams Lake Airport is closed due to low ceilings.

The Cariboo Mountains to the East of Quesnel and Williams Lake

(Not shown)

Flight routes east into the valleys of the Cariboo Mountains are common especially during the summer months. Weather systems impact somewhat harder in these areas with lower ceilings and visibility. This area is well known for its summer thunderstorm activity with the accompanying gusty winds and turbulence.

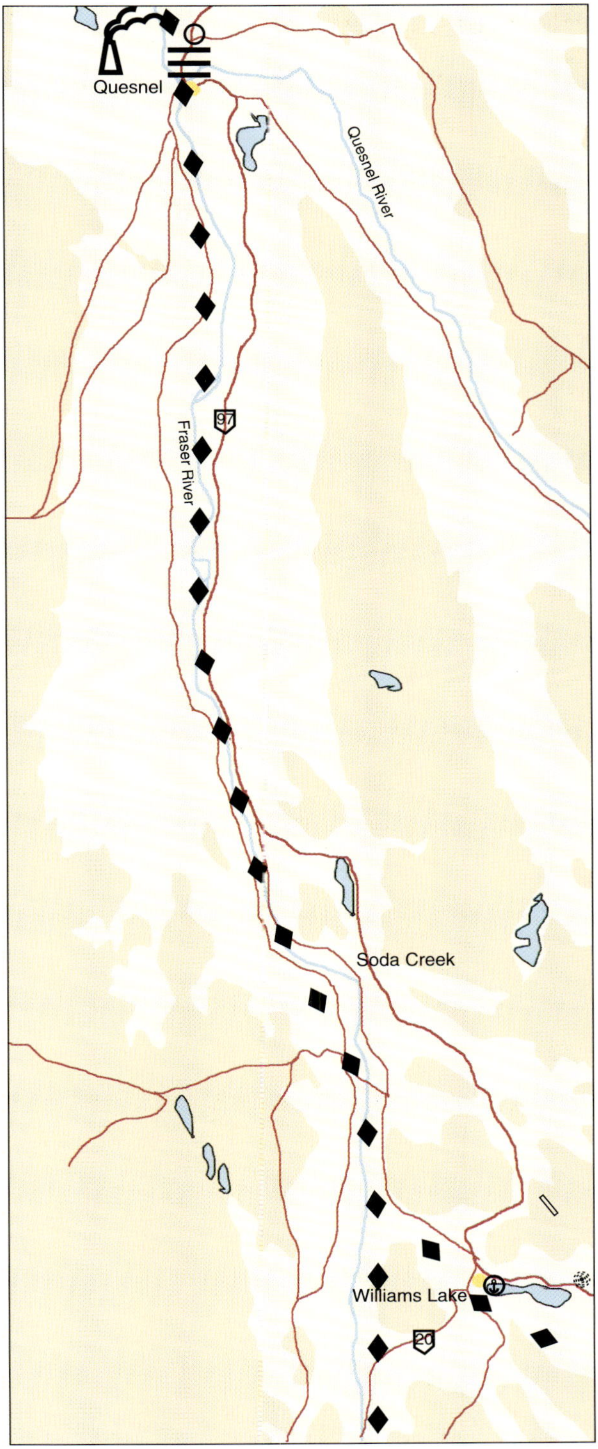

Map 29 — Quesnel to Williams Lake

(Map 30)

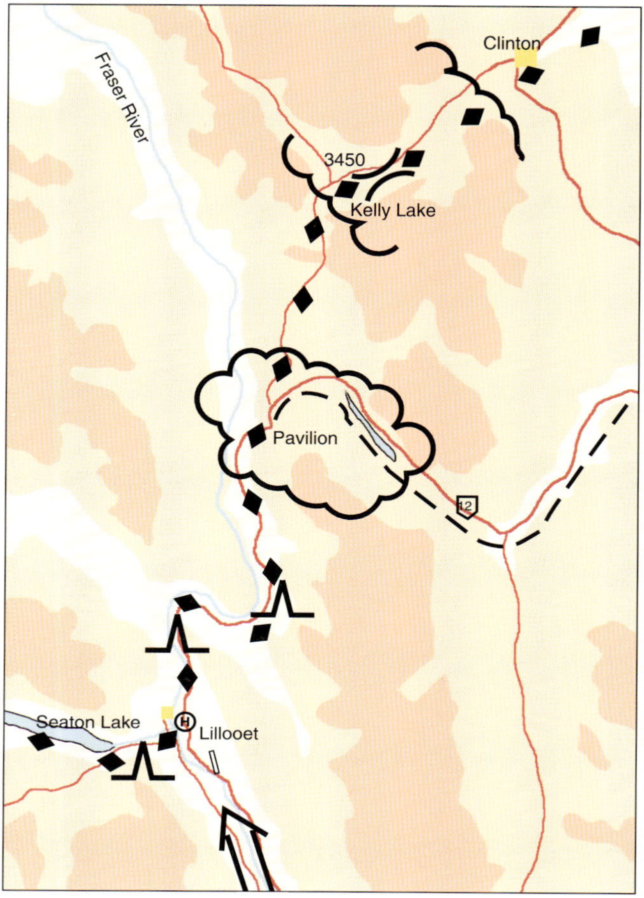

Low cloud and fog are usually concentrated over the plateau while the lower elevations such as along the Fraser River Valley are often below the cloud deck. If the ceiling is too low to get through Pavilion, it is likely the same at Kelly Lake.

Map 30 — Lillooet - Pavilion - Clinton

Cumulonimbus cloud ...

Williams Lake to 100 Mile House

(Map 31)

Conditions along this route tend to be uniform, although ceilings lower as the terrain rises when heading southeast. Ceilings are much lower over the higher ground while fog is more prevalent in the valley.

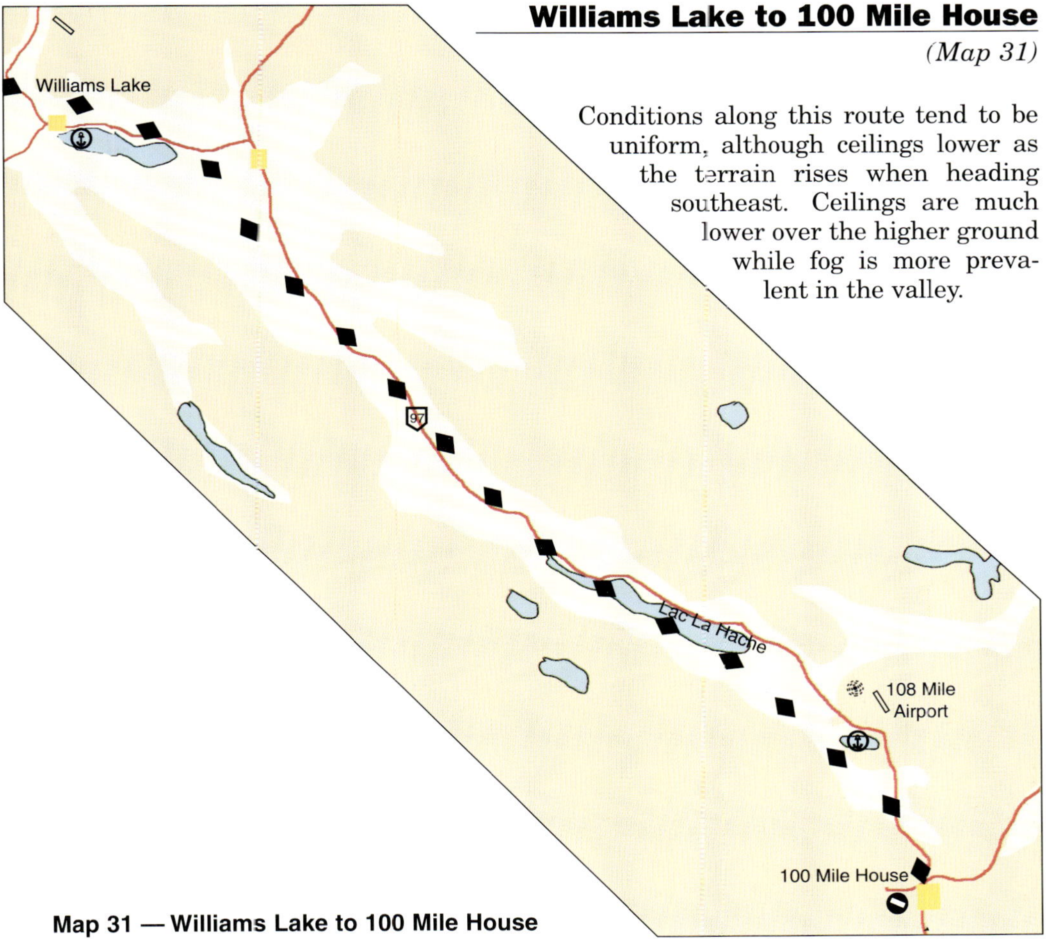

Map 31 — Williams Lake to 100 Mile House

... building northeast from Prince George Airport

T. Gairns

100 Mile House to Cache Creek

(Map 32)

Low cloud develops in the rising ground just south of Clinton and usually extends across the Central Plateau.

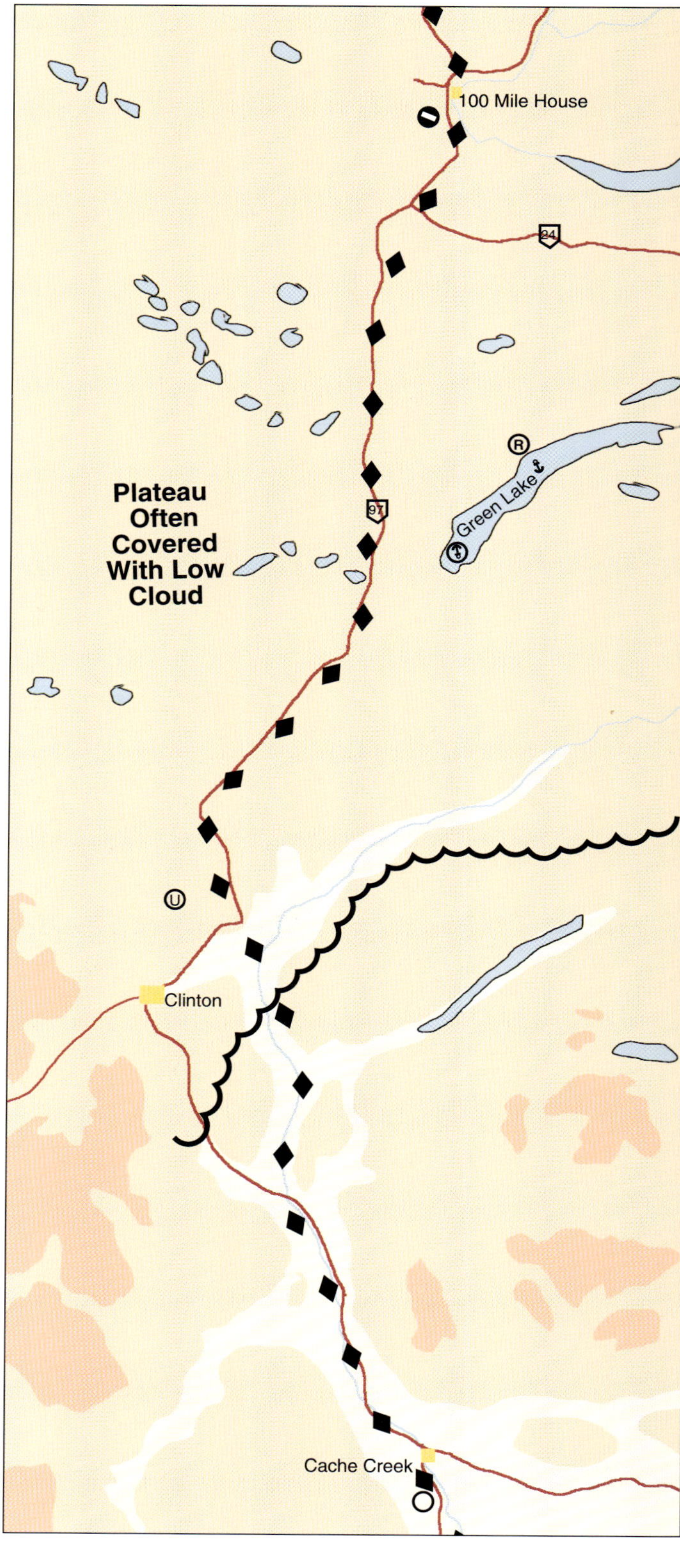

Map 32 — 100 Mile House to Cache Creek

Cache Creek to Kamloops

(Map 33)

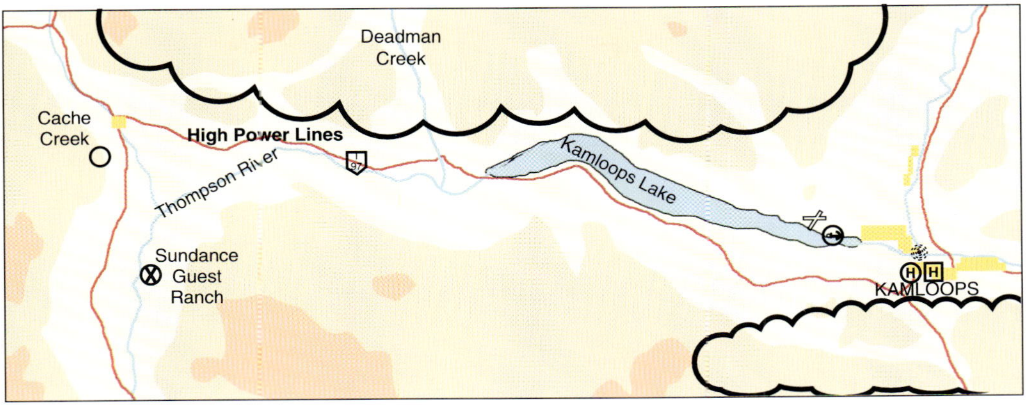

Map 33 — Cache Creek to Kamloops

Low cloud builds over Kamloops Lake but usually dissipates near Cache Creek. If passage is difficult, beware of the sucker hole at the west end of Kamloops Lake. Deadman Creek heads north and usually looks good but it narrows rapidly and rises into the plateau where ceilings will be quite low.

G. Fricska

Kamloops Lake

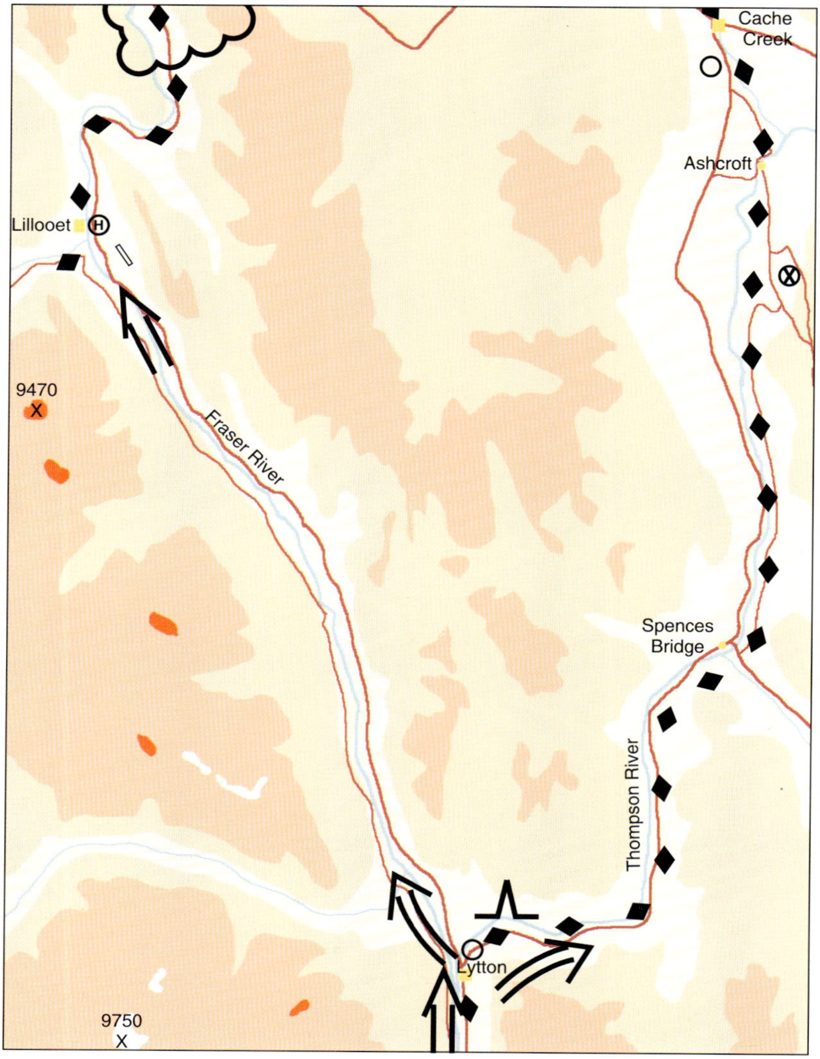

Map 34 — Lytton to Cache Creek / Lillooet

Lytton to Hope

(Maps 35 north and south)

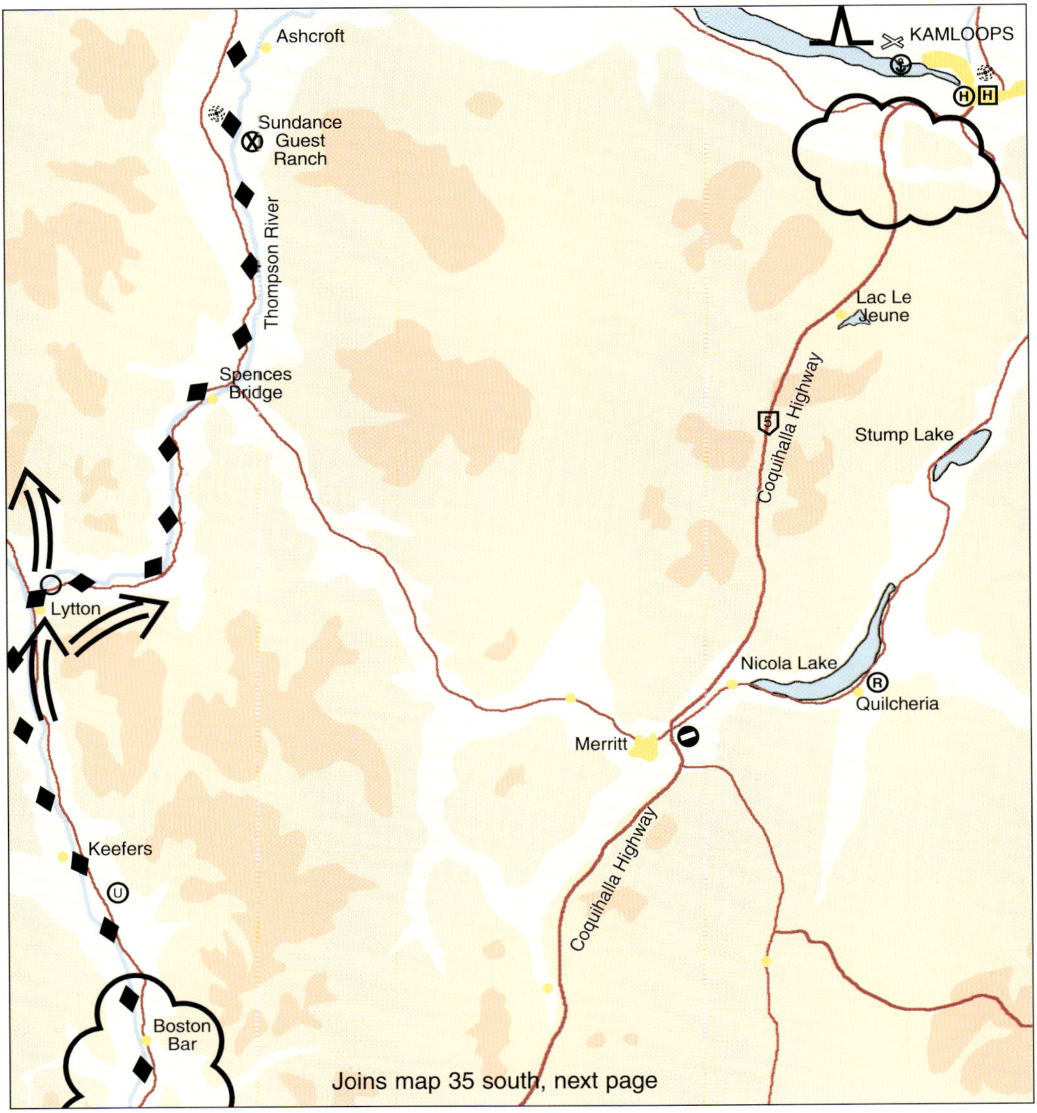

Map 35 north — Lytton to Hope.

This has long been one of the primary routes from the coast to the interior of British Columbia. If the Hope area is impassable then the Fraser Canyon as far north as Boston Bar will also likely impassable. During the fall and winter, the route from Lytton south is prone to patchy fog developing overnight and being slow to dissipate the following day. Despite frequent reports of strong winds at Lytton, turbulence is seldom encountered. The lowest conditions usually occur around Yale and improve further north. During times of strong southerly winds (30 knots or greater) severe turbulence should be expected near Lady Franklin Rock (across from the Yale Tunnel). North of Boston Bar, little turbulence has been noticed by pilots despite wind gust values of 60 knots. The area around the Lytton airport can be turbulent at times due to the splitting of the strong winds into the Fraser and Thompson River valleys.

Princeton to Hope

(Map 35 south)

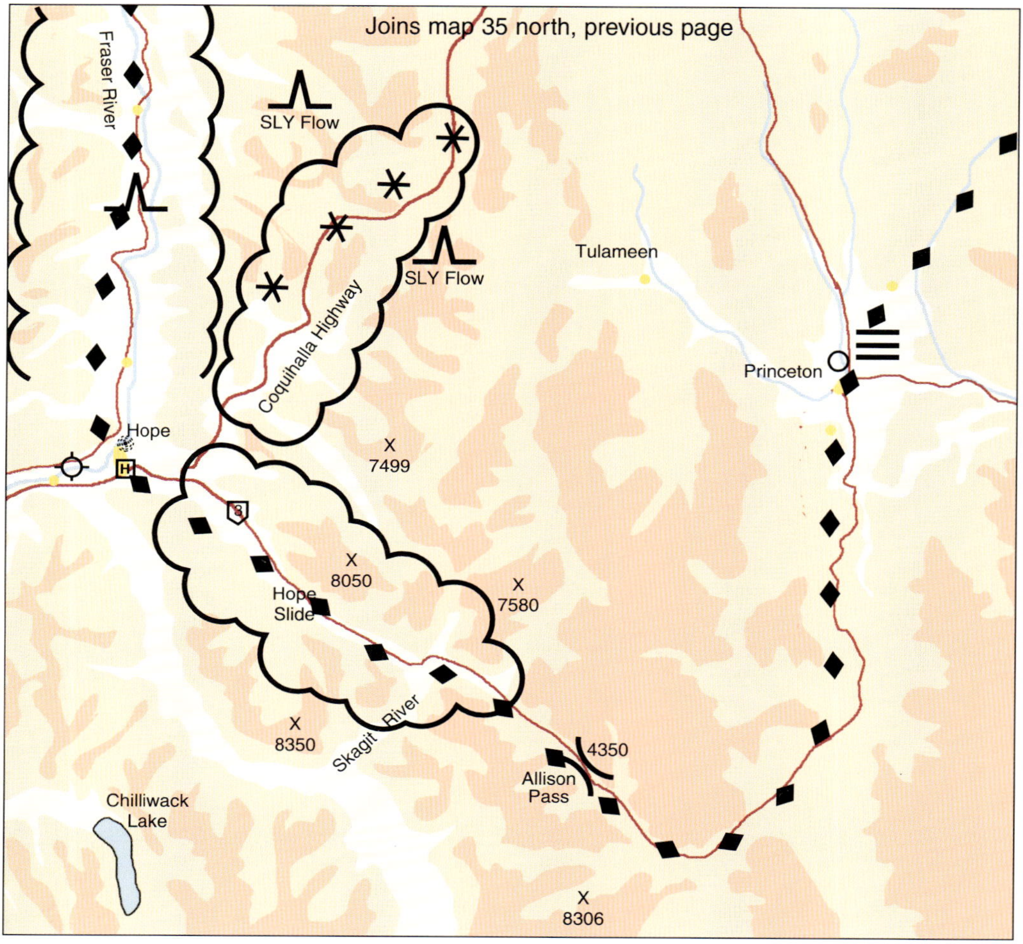

Map 35 south — Lytton to Hope.

If Hope is impassable then the Hope-Princeton west of Allison Summit is likely impassable. The Hope Slide area is particularly treacherous as the barren hillside reflects light, especially when snow-covered, causing a "brightening" in the cloud and making pilots think that conditions are improving ahead. Visibility falls rapidly along this route once steady precipitation begins. As little as ten minutes can see the route change from passable to closed. When traveling westbound, low cloud will begin to force aircraft to descend just east of the Skagit Valley but it still remains that visibility will be the main problem.

When traveling eastbound from Hope to Princeton, experts recommend that aircraft pass Hope with sufficient altitude to safely cross Allison Pass. East of the Skagit River, the valley rises very steeply to the summit. This gradient exceeds the climb capability of most conventional aircraft and the rapidly narrowing valley makes safe turning dangerous or impossible.

Cache Creek - Lillooet - Pemberton, Highway 97

(Map 36)

This is a popular route to fly to the coast often avoiding poor conditions and strong winds occurring in the Fraser Canyon. However, low cloud is common near Pavilion. Occasionally in summer, strong thunderstorms can be encountered and in fall fog can be a problem near the lakes.

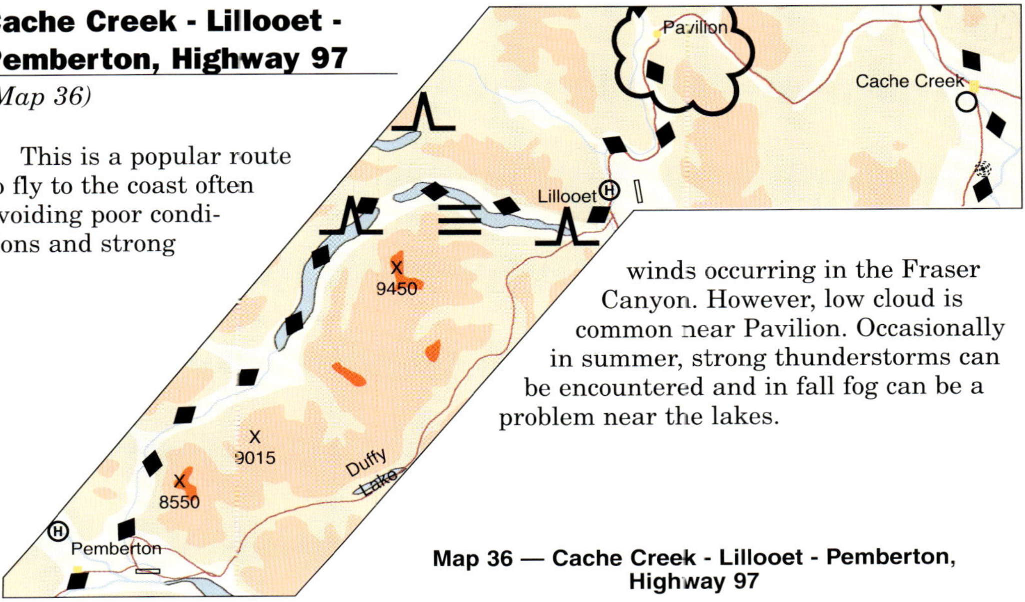

Map 36 — Cache Creek - Lillooet - Pemberton, Highway 97

Williams Lake to Alexis Creek

(Map 37)

Low cloud and fog are common in the fall and winter season especially at the lower ele-

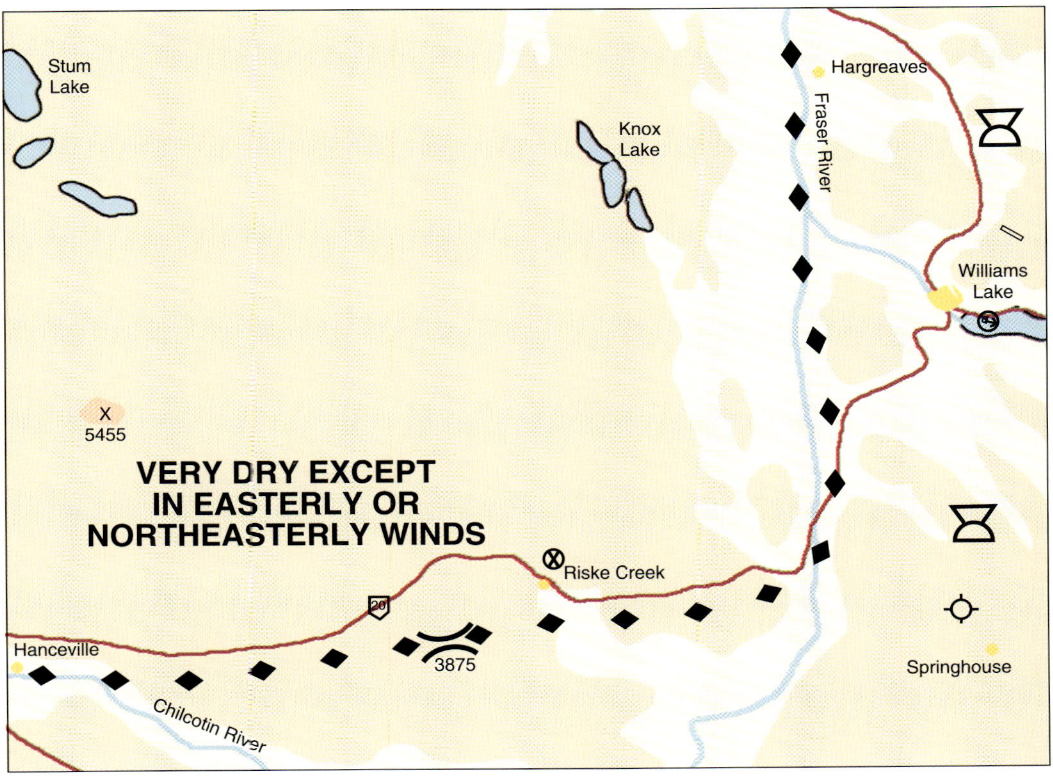

Map 37 — Williams Lake to Alexis Creek

vations. The fog and low cloud are more likely to burn off by late morning near Alexis Creek.

Light to moderate turbulence usually occurs during the afternoons in the summer season due to convective heating. Thunderstorms are more frequent along the Fraser and eastward with most cells smaller and in the developmental stage west of the Fraser.

Generally, weather systems give more precipitation east of the Fraser with drier conditions prevailing west of the Fraser.

Alexis Creek to Puntzi Mountain

(Map 38)

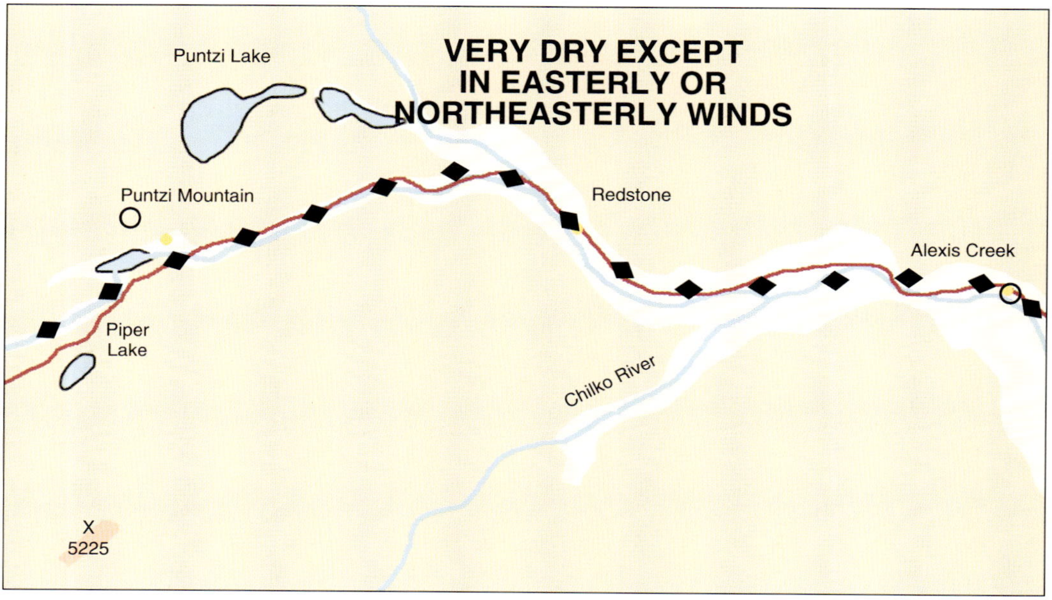

Map 38 — Alexis Creek to Puntzi Mountain

Low cloud and fog can be a problem in the fall and winter season especially near the larger lakes until they freeze over. Low cloud can occur with a moist easterly circulation at any time of year.

Chilko Lake *G. Fricska*

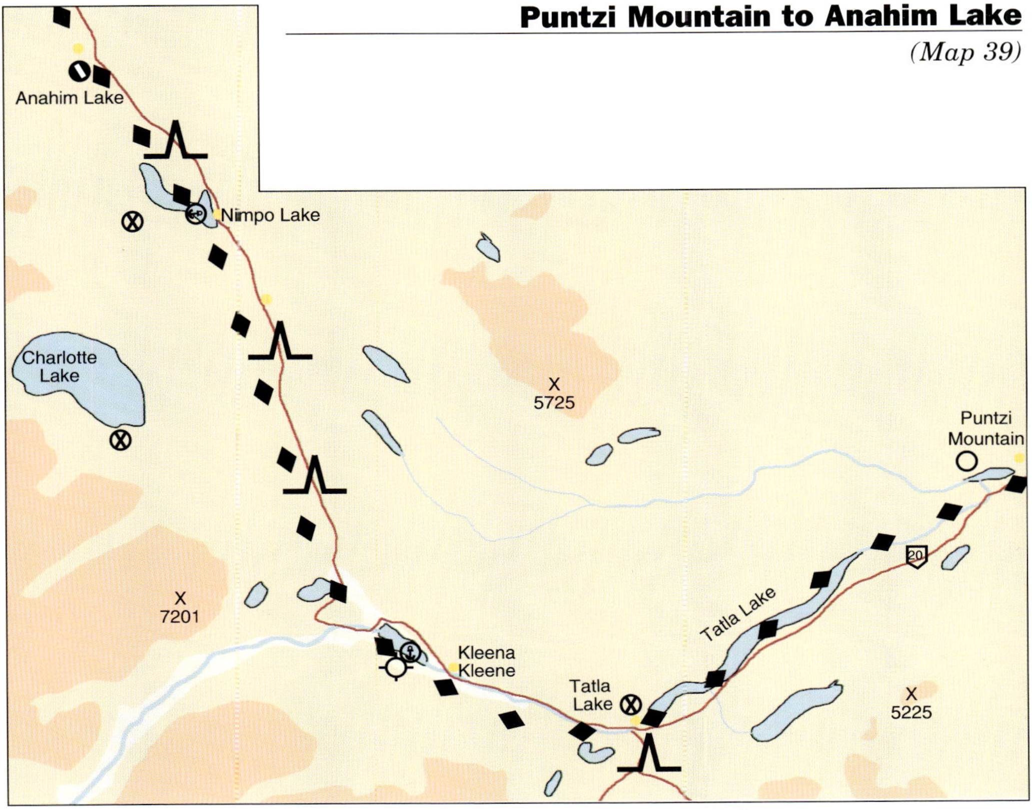

Puntzi Mountain to Anahim Lake

(Map 39)

Map 39 — Puntzi Mountain to Anahim Lake

This region is dry belt country. Severe turbulence is common along with very strong winds during the summer to the east of the Coast Mountains. This includes Tatla and Chilko Lakes, among others. In most cases, to avoid this turbulence, it would be necessary to ascend to nearly 15,000 feet. Experienced pilots suggest that it is best to stick close to the sides of the valleys.

Anahim Lake to Bella Coola

(Map 40)

This route goes from predominantly dry belt weather to coastal weather in a fairly short distance. Low cloud is usually encountered from the crest of the Coast Mountains westward to Bella Coola. The highway is not the route of choice due to its higher elevation. The

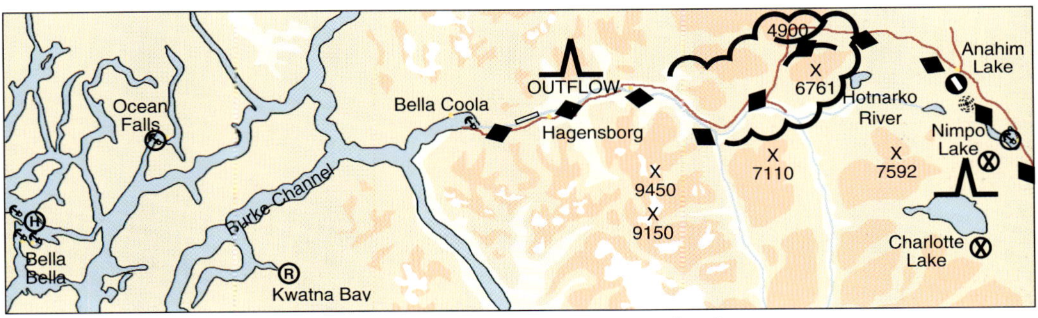

Map 40 — Anahim Lake to Bella Coola

preferred route is to follow the Hotnarko and Atnarko river valleys to the Bella Coola Valley due to lower elevations. The Bella Coola Valley frequently has fog or low cloud which can persist all day during the fall and winter seasons.

In the winter, high pressure will often dominate over the interior giving outflow conditions. This can produce significant downslope winds and turbulence over eastern sections of the Bella Coola Valley.

Watson Lake to Dease Lake

(Map 41)

Low cloud and fog are common over Dease Lake in autumn. Heavier precipitation and more frequent cloud are generally experienced near Cassiar. Most VFR traffic follows the much wider Dease River valley, rather than the highway.

South winds tend to be strong all summer at Dease Lake, producing some turbulence near the airport.

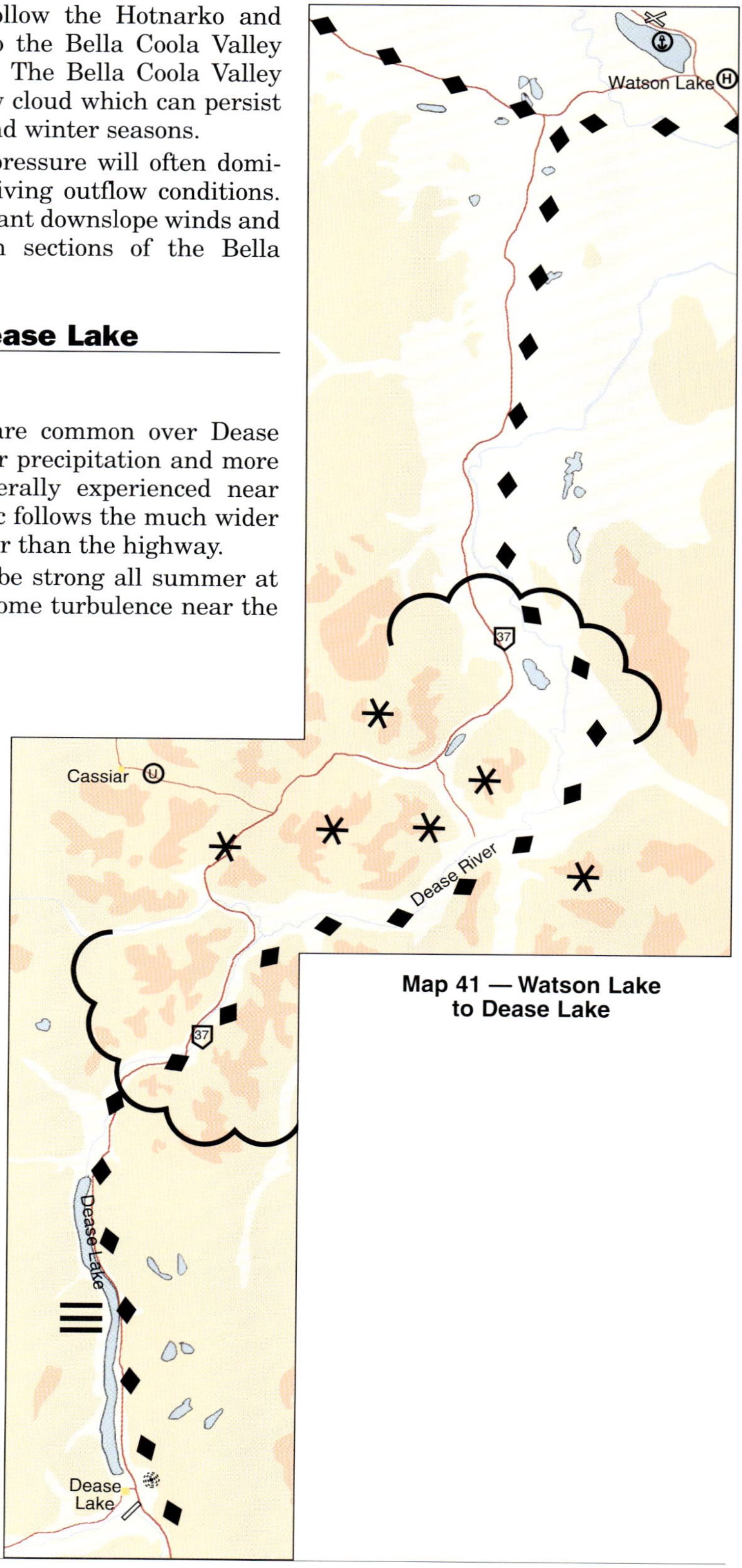

Map 41 — Watson Lake to Dease Lake

Dease Lake to Meziadin

(Map 42 and 43)

The high pass just south of Dease Lake can be closed by low cloud with little or no cloud in the surrounding areas. Low cloud often closes the route just north of the Meziadin Junction as well as in Ningunsaw Pass south of Bob Quinn Lake. The Pass is also prone to heavy snow in winter as the weather funnels up from the coast. The low cloud tends to envelop this area before it reaches Smithers or even Terrace. Weather observation at Stewart should be examined for an idea of conditions near Meziadin. Inflow up the Iskut Valley often leads to upslope low cloud south of Burrage Creek.

The side valley from Meziadin Lake to Stewart is spectacular but, unfortunately, often unsuitable for VFR flying *(see also map 44, page 105)*.

> ***Weather Hint:***
>
> Ketchikan Airport can give some idea of the weather in the region of Meziadin. If there is low cloud at Ketchikan along with a westerly wind component, the route is likely closed.

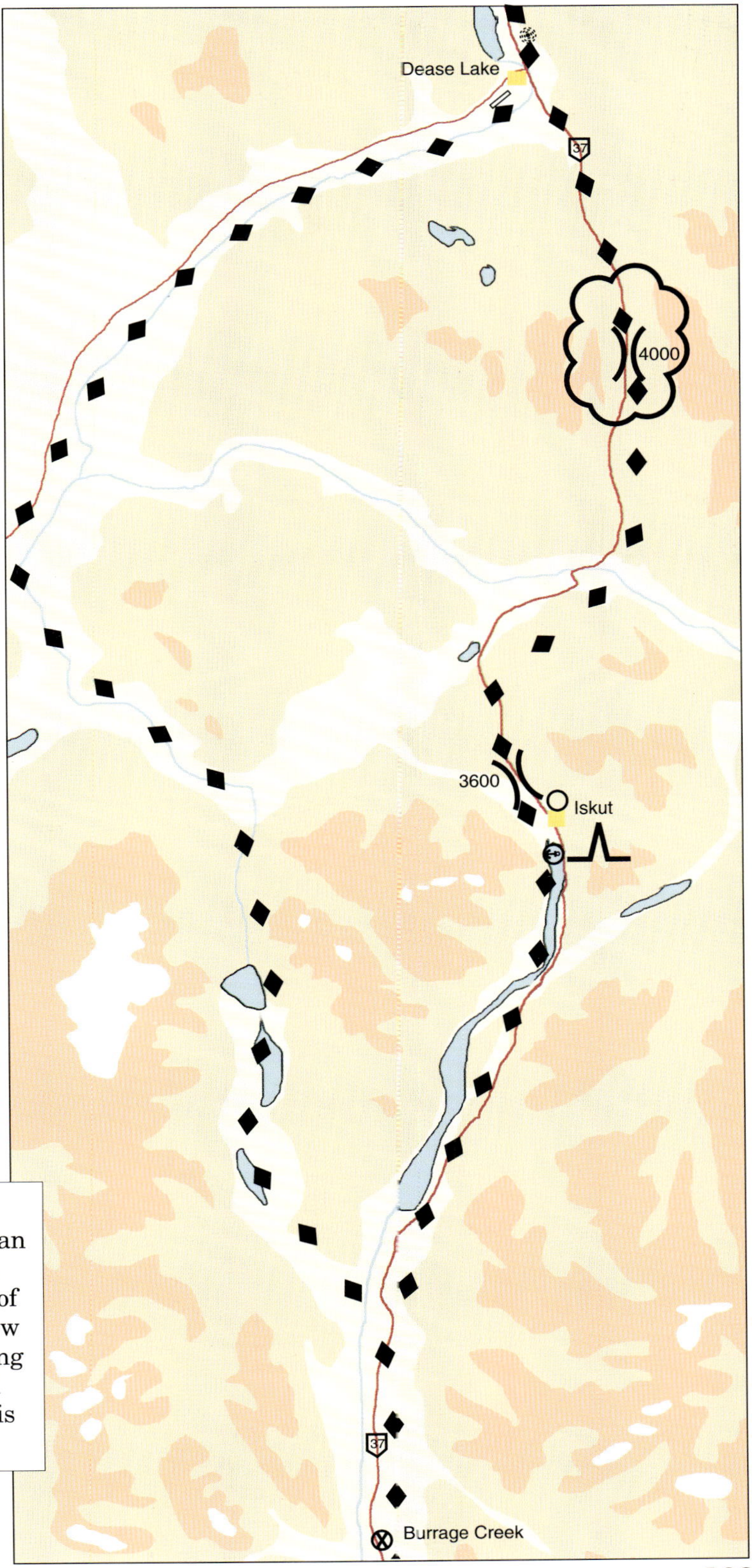

Map 42 — Dease Lake to Burrage Creek

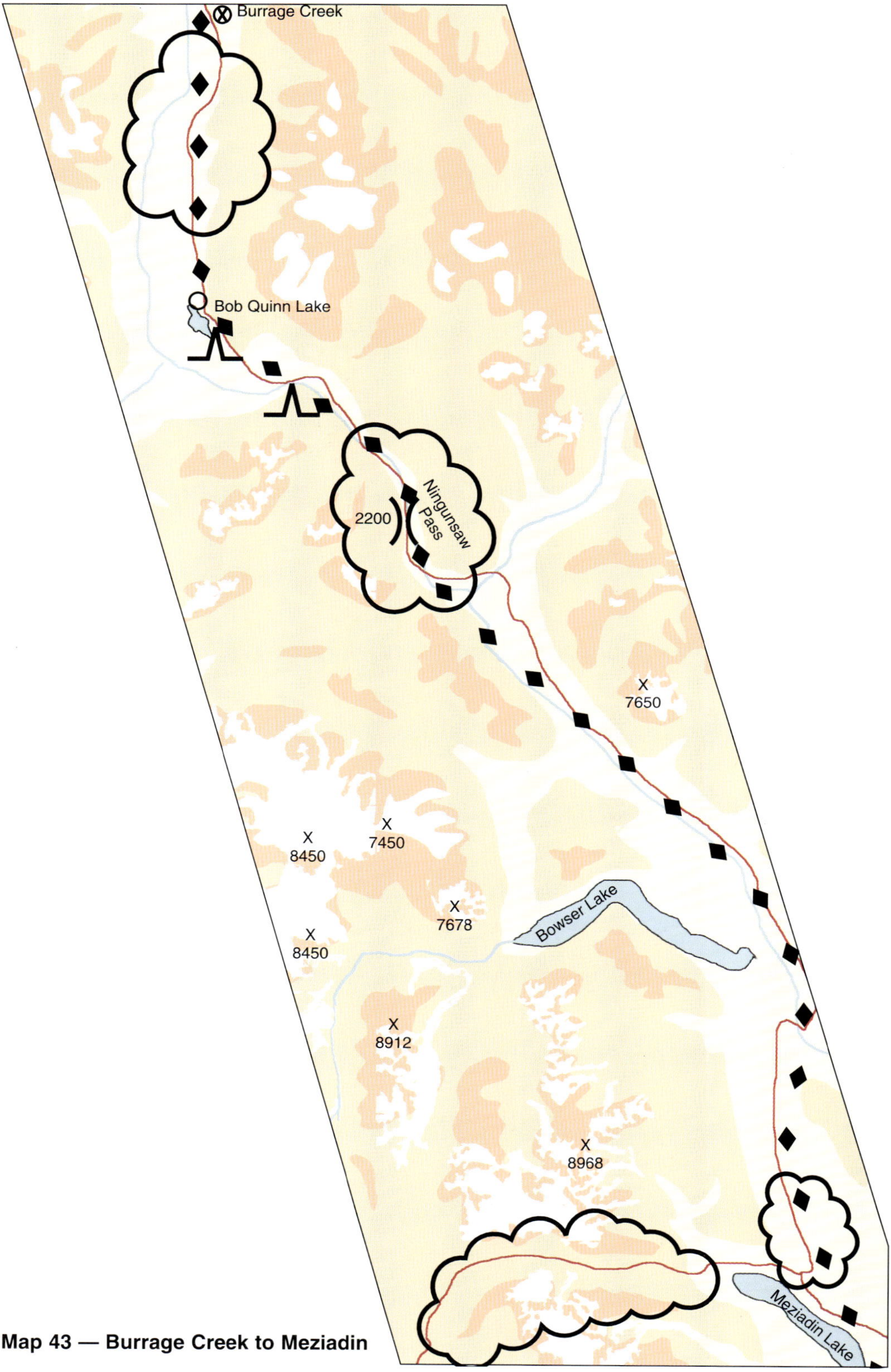

Map 43 — Burrage Creek to Meziadin

Meziadin Lake to Hazelton

(Map 44)

The weather along the highway and Kispiox River is relatively uniform both topographically and meteorologically.

Only one location has been singled out as more prone to low cloud, that being just south of Kitwancool Lake, in a southerly upslope flow. It is flat and swampy in places with frequent invasions of Pacific moisture up the Nass Valley. Radiation fog is common in autumn.

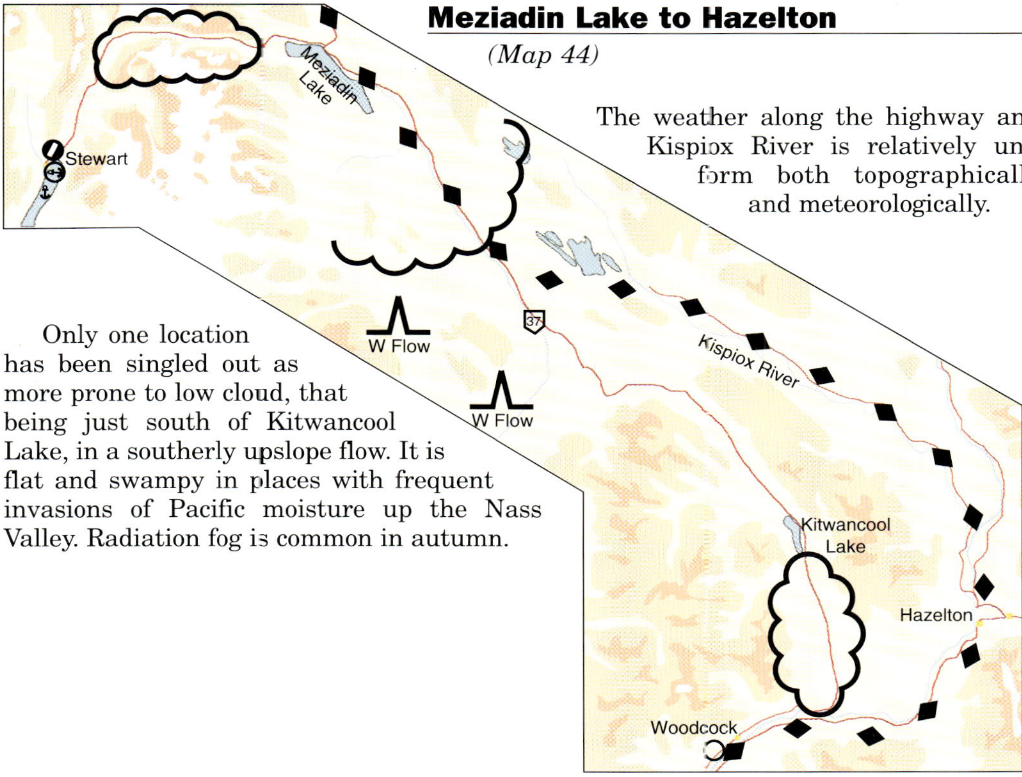

Map 44 — Meziadin to Hazelton

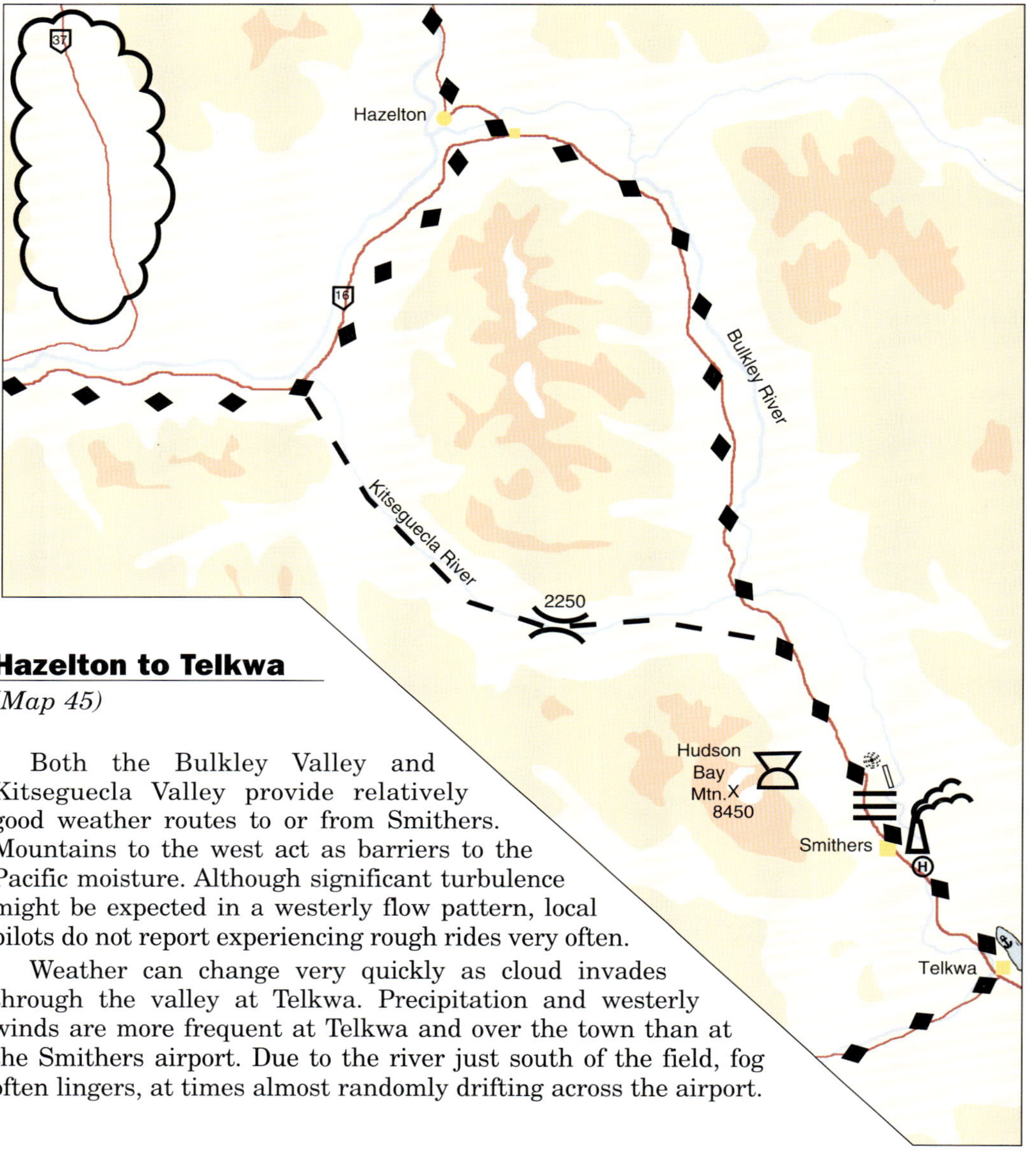

Hazelton to Telkwa

(Map 45)

Both the Bulkley Valley and Kitseguecla Valley provide relatively good weather routes to or from Smithers. Mountains to the west act as barriers to the Pacific moisture. Although significant turbulence might be expected in a westerly flow pattern, local pilots do not report experiencing rough rides very often.

Weather can change very quickly as cloud invades through the valley at Telkwa. Precipitation and westerly winds are more frequent at Telkwa and over the town than at the Smithers airport. Due to the river just south of the field, fog often lingers, at times almost randomly drifting across the airport.

Map 45 — Hazelton to Telkwa

Hazelton to Terrace

(Map 46)

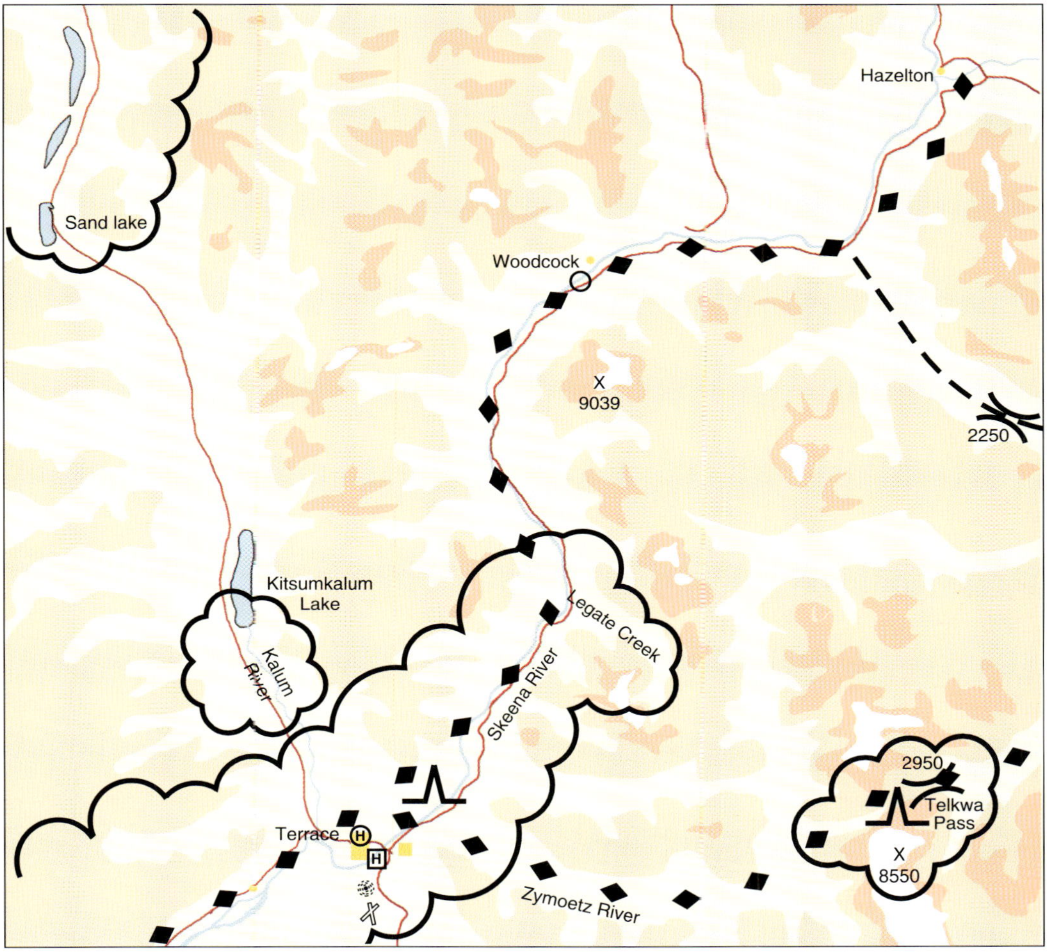

Map 46 — Hazelton to Terrace to the Nass Valley

There is a significant change in the precipitation regime along the Skeena River. Typical coastal forests change to drier interior varieties just south of Woodcock. Low cloud often extends up the Skeena, with Legate Creek a known bad spot. The Terrace Airport weather is generally representative of that up the valley. Intrusions of low cloud and fog with inflow winds are the main weather hazard and, although more frequent in winter, can occur throughout the year. This route is sometimes drier and smoother than the Telkwa pass option. Both options can deteriorate quickly near Terrace. When winds aloft at 6,000 feet exceed 30 knots, moderate turbulence is common throughout the region. Rounding Copper Mountain at the confluence of the Copper (Zymoetz) and Skeena rivers, there is often a wall of wind giving sudden and unexpected turbulence. This usually occurs when strong inflow winds are experienced at the Terrace Airport.

Terrace to the Nass Valley

(Map 46, above)

If even a scattered layer of low cloud below 700 feet is reported with a light southerly wind at the Terrace Airport, the Kalum River Valley is likely socked in. Low cloud tends to plug the route from Terrace to the Nass Valley, just north of Sand Lake, in a light southerly flow. With inflow up the Nass Valley, worst conditions can be expected in an upslope area along the road just south of the junction.

Telkwa to Terrace

(Map 47)

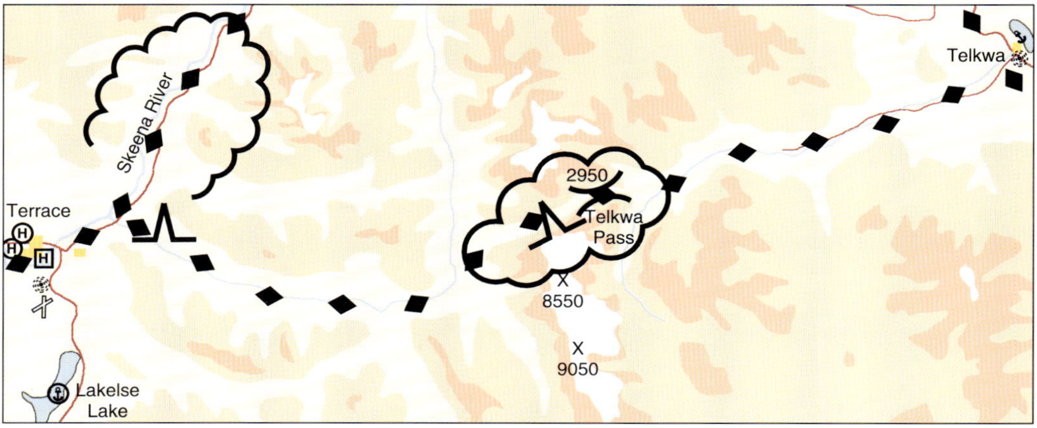

Map 47 — Telkwa to Terrace

The route through Telkwa pass is another of B.C.'s masterpieces during good weather with hanging glaciers on both sides of the valley. When westbound through Telkwa Pass with low cloud or freezing levels, sudden snow squalls or rapidly lowering cloud can reduce ceiling and visibility to near zero almost instantly. The low cloud or snow will often make dead-end passes appear to be the main route. This pass should be avoided in any weather conditions likely to obscure the route. Poor conditions in the pass are often not observed at either Telkwa or Terrace. Severe turbulence is reported to occur during inflow conditions along the ridge just east of the town of Terrace.

Telkwa to Burns Lake

(Map 48)

This route becomes drier as one moves from Smithers toward Burns Lake.

Convective cloud in the summer often develops and is more frequent near Burns Lake. Gusty westerly crosswinds often make landings tricky at Burns Lake.

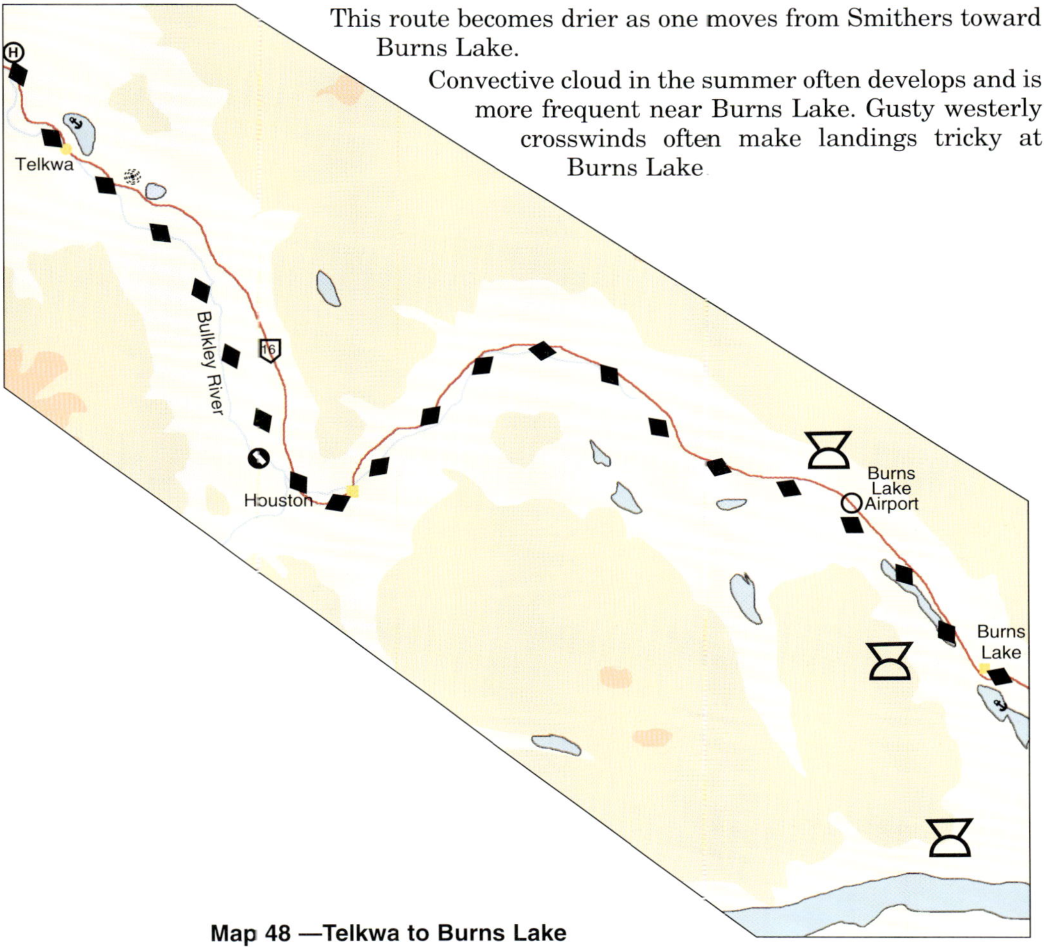

Map 48 —Telkwa to Burns Lake

G. Fricska

Western interior of B.C.

Burns Lake to Vanderhoof

(Map 49)

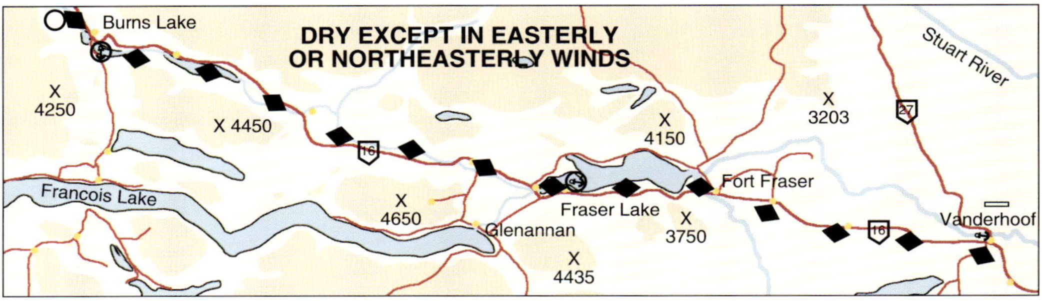

Map 49 — Burns Lake to Vanderhoof

Vanderhoof to Prince George

(Map 50)

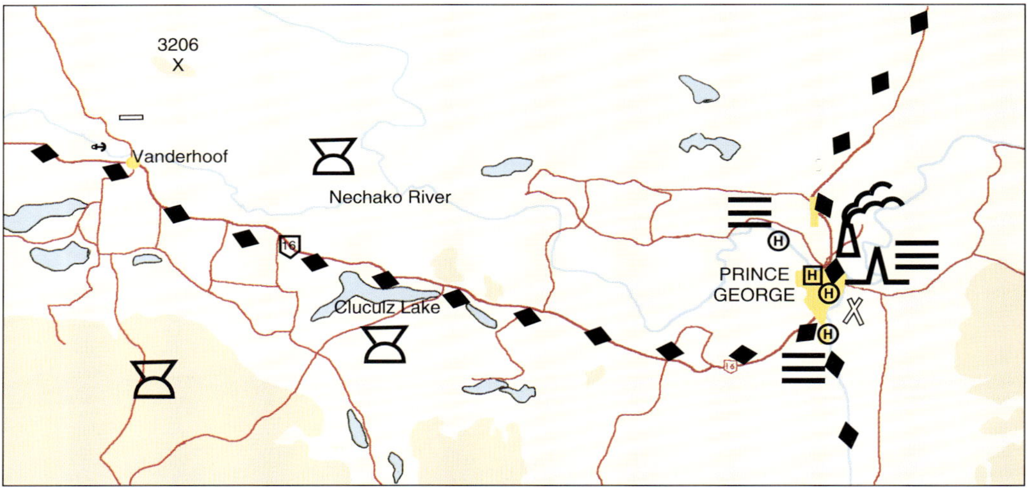

Map 50 — Vanderhoof to Prince George

(Map 51)

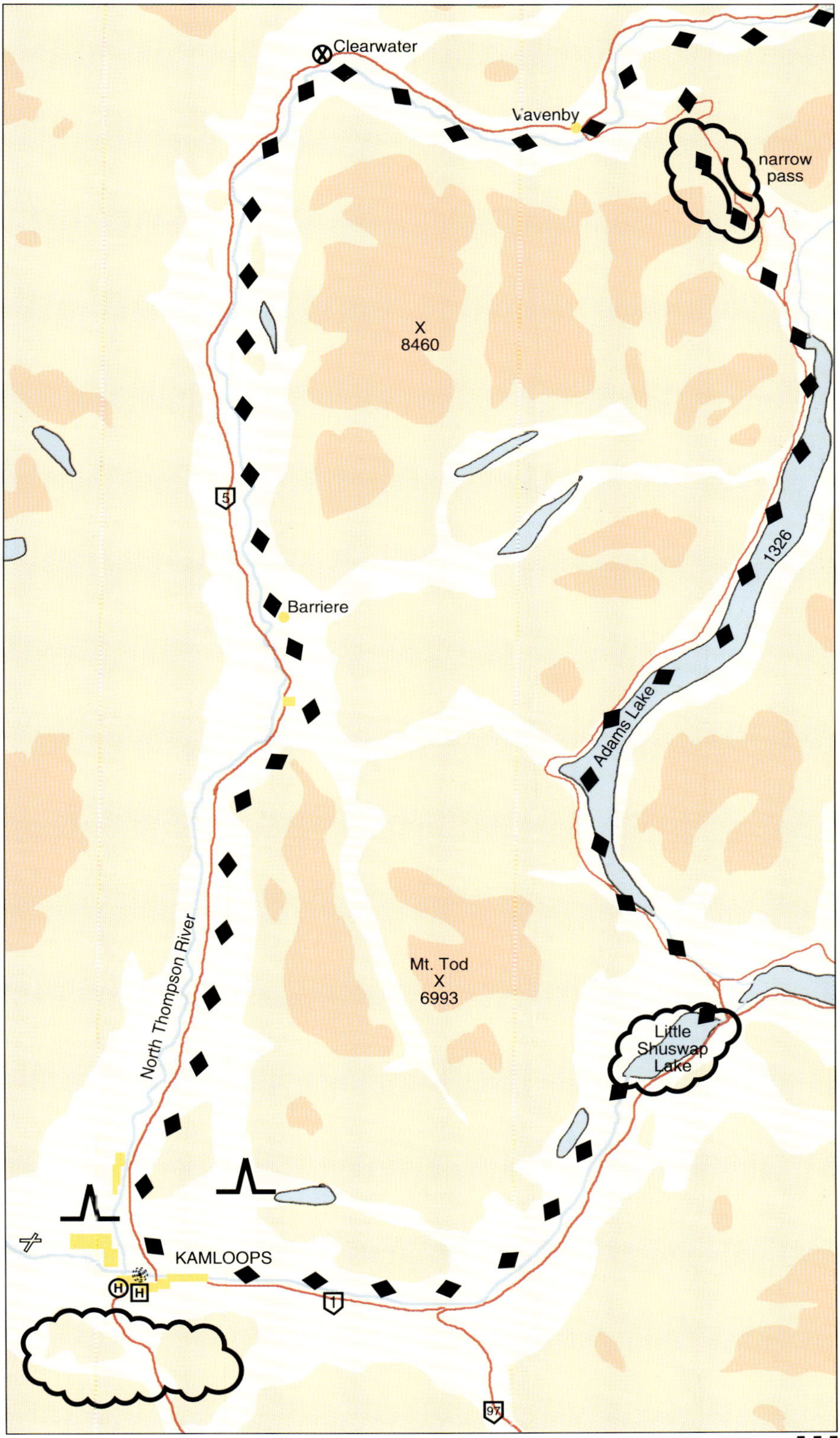

Map 51 — Kamloops to Vavenby

Two routes are commonly used between Kamloops and Vavenby; one along the highway and one along the lake. Just to the north of Adams Lake the combination of low cloud over rising ground and narrow passes can be quite hazardous. Depending on the moisture and the time of year, ceilings can be right down onto the lake. It should be noted that Adams Lake remains open all year.

Vavenby to Blue River

(Map 52)

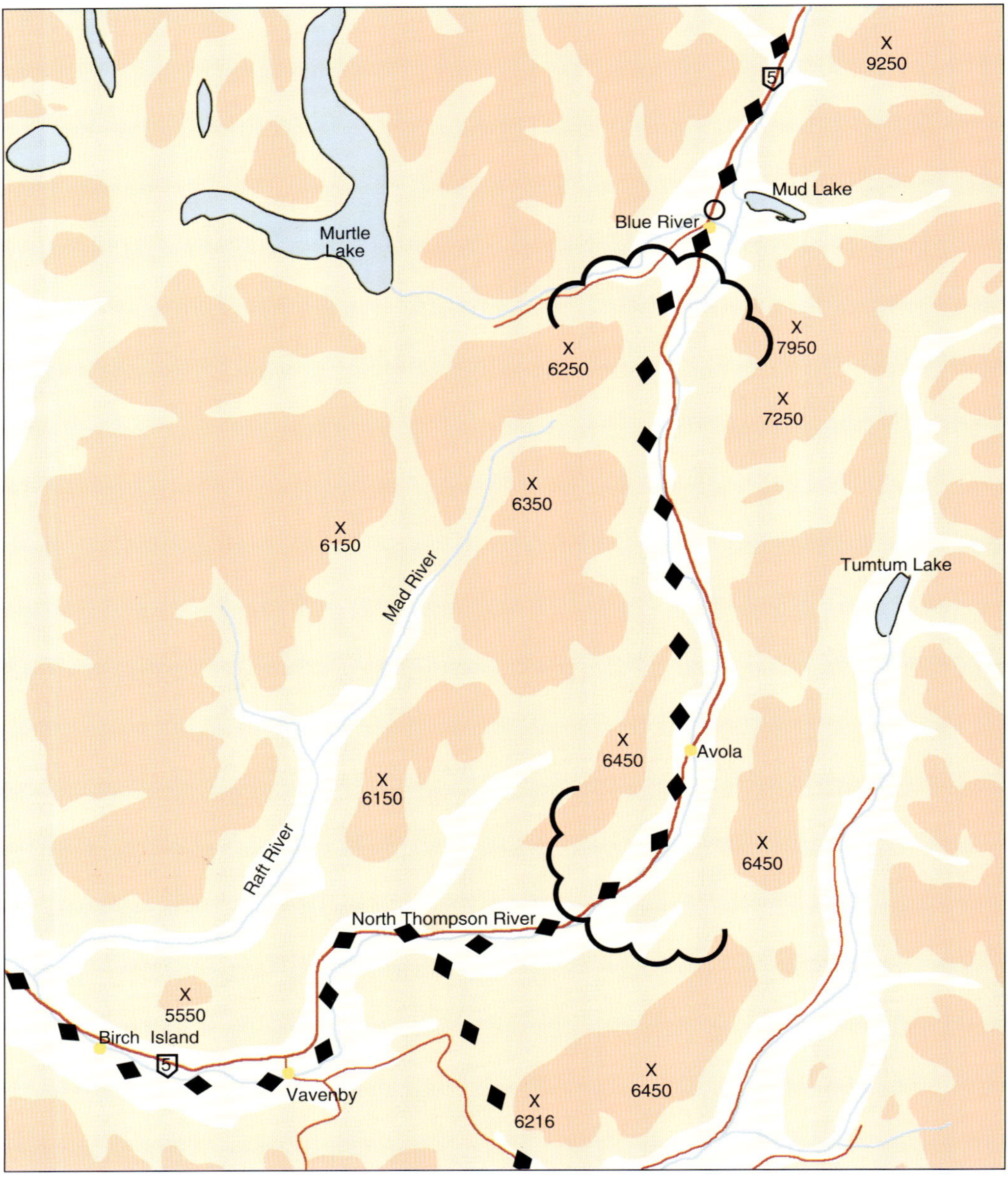

Map 52 — Vavenby to Blue River

Blue River weather observation does not usually report the low cloud that forms over the rising ground just to the south.

Blue River to Tete Jaune

(Map 53)

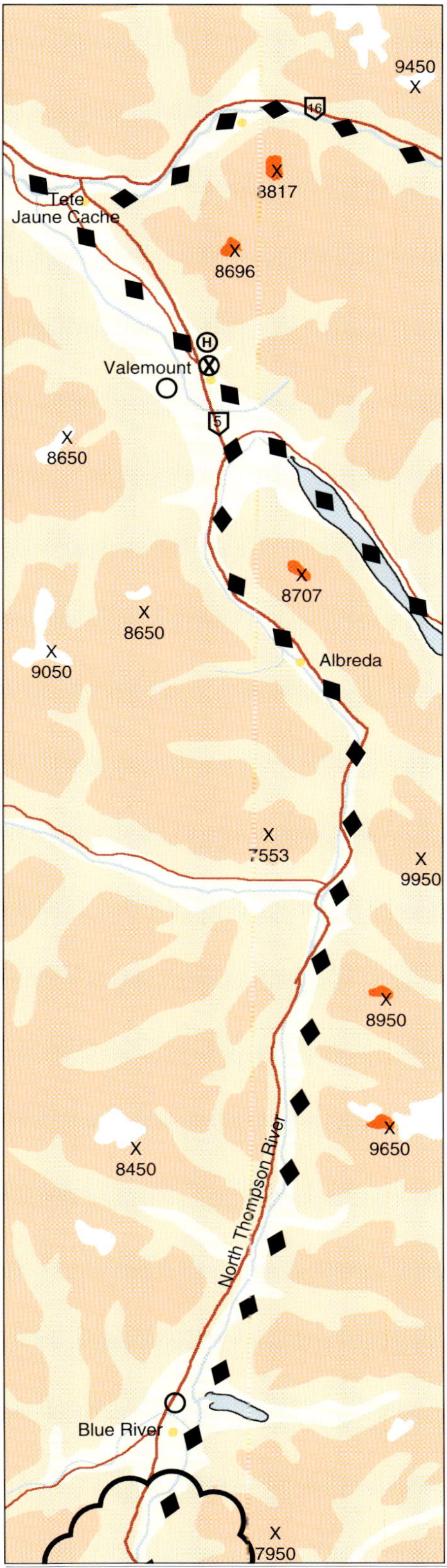

Map 53 — Blue River to Tete Jaune

Kamloops to Salmon Arm

(Map 54)

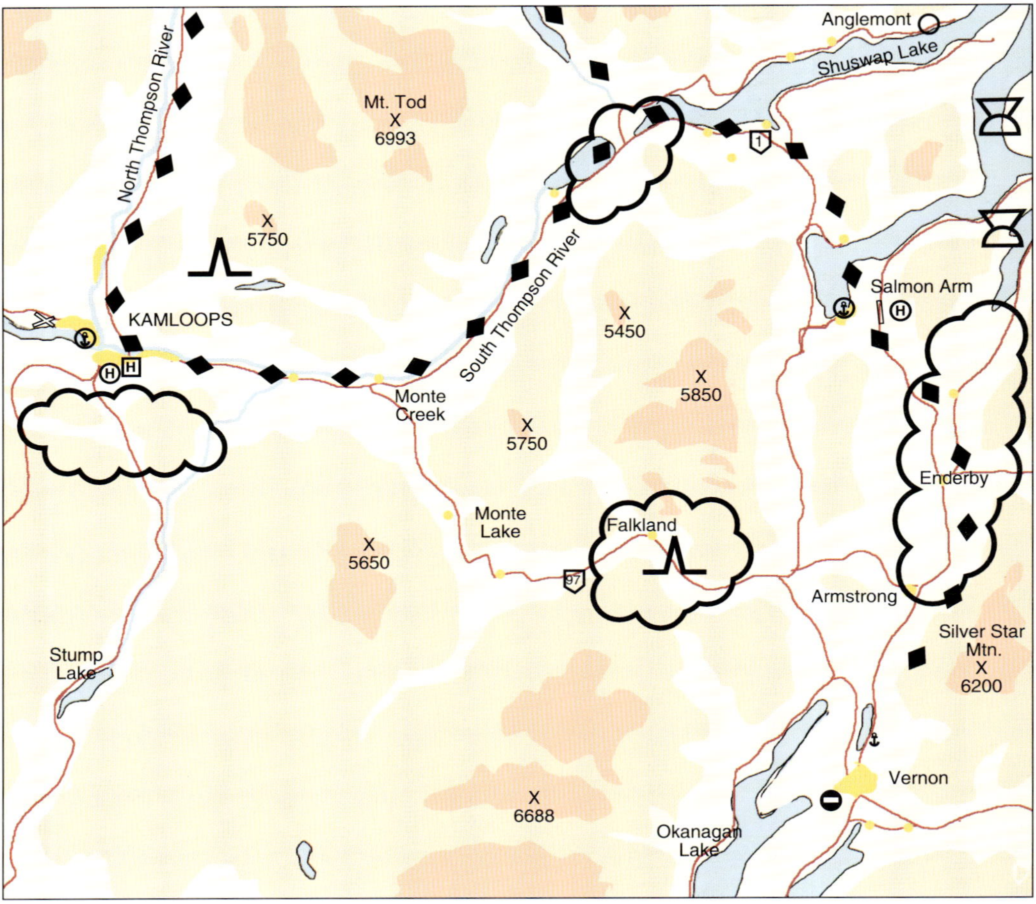

Map 54 — Kamloops to Salmon Arm

It can be quite turbulent at times in the wide part of the valley just to the east of Kamloops.

As you approach Shuswap Lake, low cloud develops as the moisture increases and is quite common from October to mid-March. It is also worth noting that the arms of Shuswap Lake freeze over but the deep central part usually remains open.

Vicinity Salmon Arm

(Map 54, above)

The airport is 600 feet higher than Shuswap Lake and can make approaches difficult during times of prolonged low valley cloud.

During the summer months, thunderstorms tend to intensify near Shuswap Lake and many feel that the lightning capital of B.C. is the area just to the east of the lake.

Salmon Arm to Revelstoke

(Map 55)

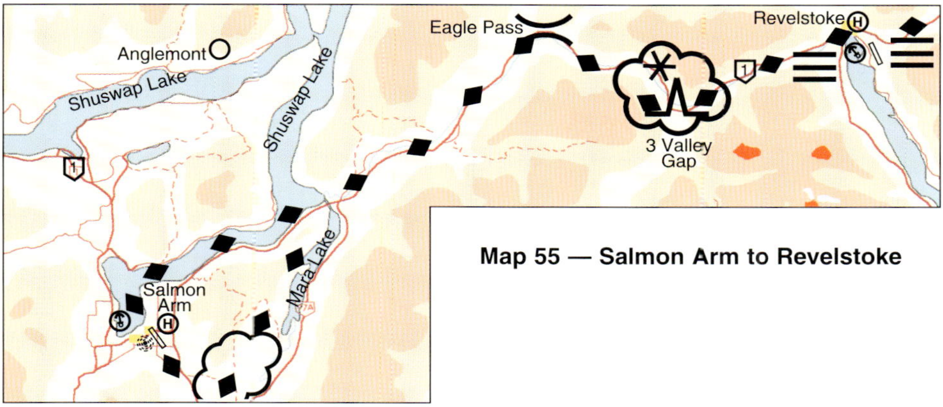

Map 55 — Salmon Arm to Revelstoke

If you can get out of Revelstoke, you can usually get to Salmon Arm. The main hazard along this route is 3 Valley Gap, where the valleys narrow and merge, producing low cloud, heavy snow and unpredictable winds.

Revelstoke Airport with weather station *K. Johnson*

Revelstoke Area including north to Mica Dam

(Map 56)

Because of the open water and open flats south of town, fog is common from the KOA Campground (to the east) to 3 Valley Gap (to the west), although the most common low visibility is in the flats near the airport. Some mornings, the entire region can be clear while fog closes both approaches to the Revelstoke Airport. Surface winds at Revelstoke, although they can be strong from the south, are not usually a hazard to aviation.

The route from Revelstoke north to Mica Dam is used most commonly by helicopters. The route is essentially two 700 foot steps with ceilings dropping as one travels north. Pilots must be aware that low cloud is common over open water, and beware of the power lines crossing the route several times north of Downie Creek.

> *Weather Hint:* Revelstoke airport is a good indicator of surrounding passes.
>
> If Revelstoke has low cloud, ceiling less than 3,000 feet, the passes are invariably closed.

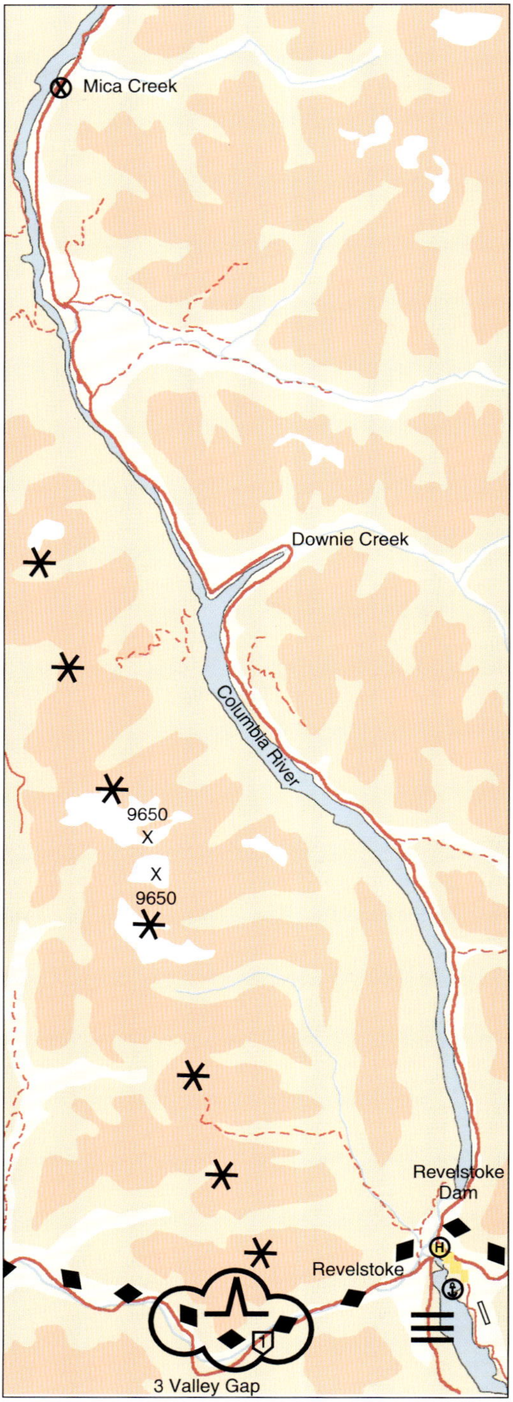

Map 56 — Revelstoke area including north to Mica Dam

Revelstoke to Golden

(Map 57)

Although a relatively wide mountain pass, Rogers Pass has intimidated even experienced pilots. The weather often looks good entering Rogers Pass from the east; however, getting to Rogers Pass summit does not assure passage to Revelstoke. Low cloud tends to pile up west of the summit of Rogers Pass.

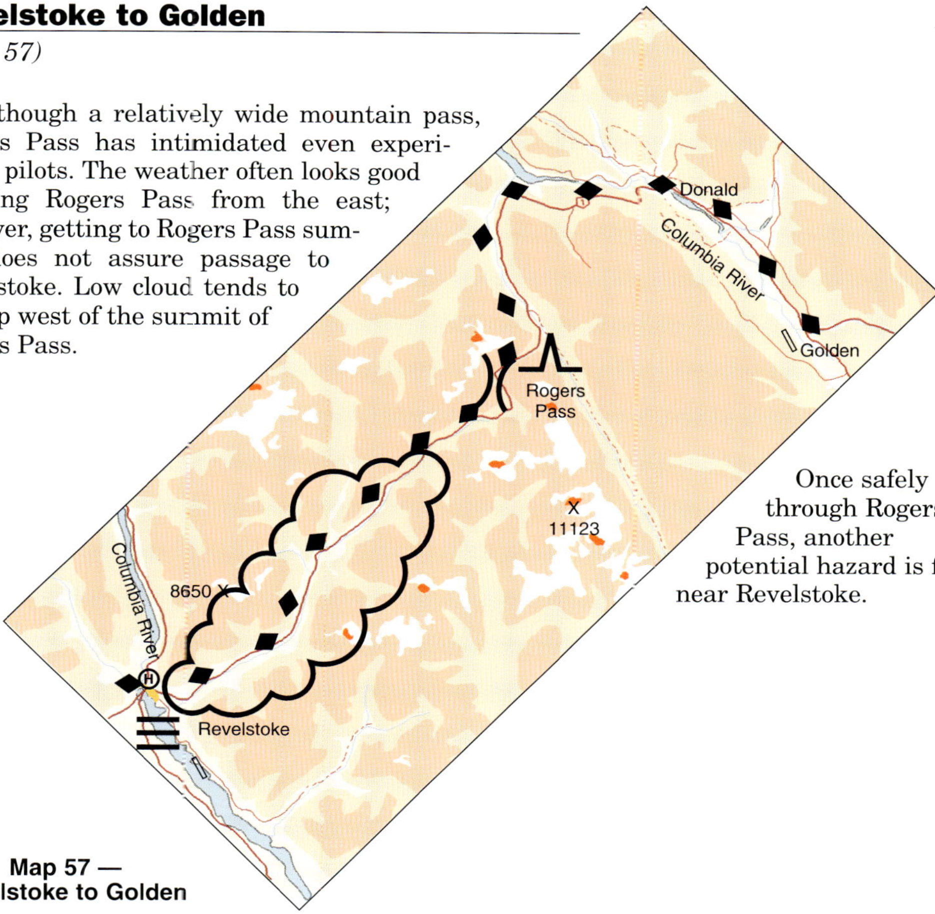

Once safely through Rogers Pass, another potential hazard is fog near Revelstoke.

Map 57 — Revelstoke to Golden

Rocky Mountains Banff to Golden

(Maps 58 and 59)

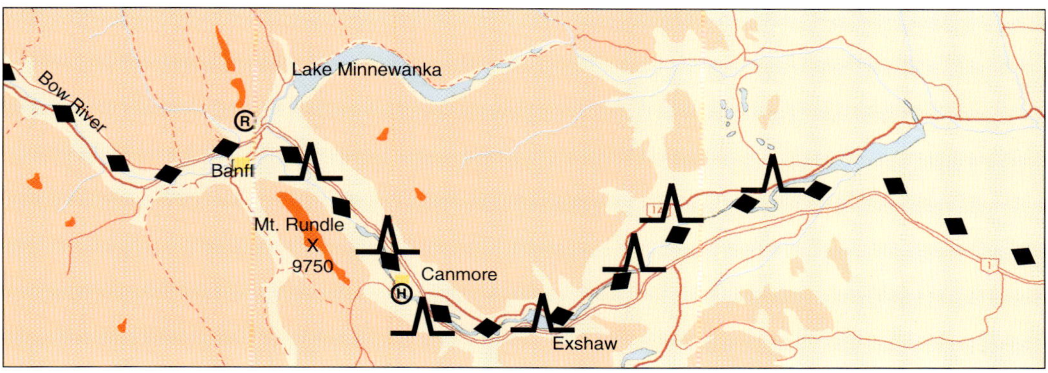

Map 58 — Rocky Mountains, Banff to Golden

The VFR route from Calgary to Golden crosses a major hydrological regime change over the Continental Divide. Differences in aviation weather along this route can be equally as dramatic. Due to typically very strong surface winds, low level turbulence is often

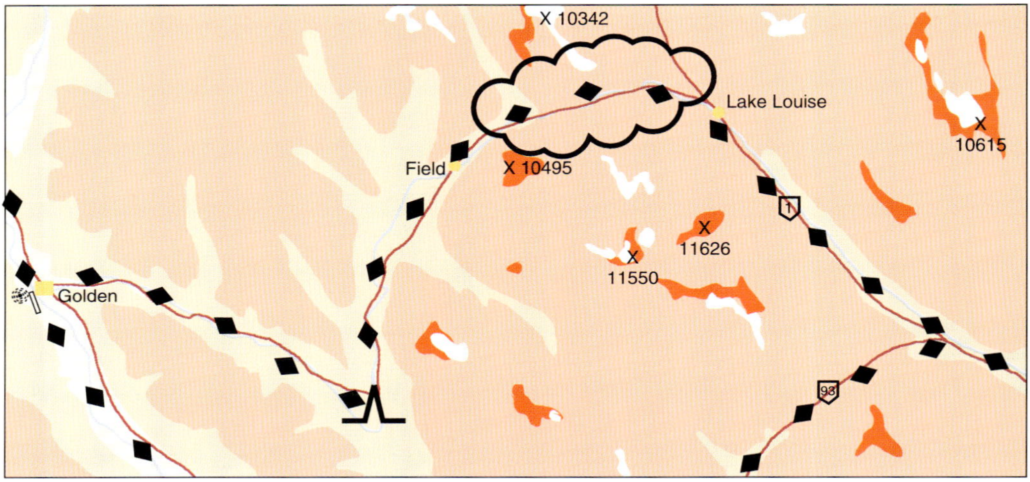

Map 59 — Rocky Mountains, Banff to Golden

moderate or greater, at times making an otherwise clear day unflyable. These strong winds almost always dissipate west of Banff. Higher altitudes, say 10,000 to 16,000 feet often provide no refuge from the turbulence due to the frequency of lee wave activity. As with the low level turbulence, this lee wave turbulence is rarely observed west of Banff.

Kicking Horse Pass near Field *K. Johnson*

Blaeberry Pass, North Saskatchewan Crossing to Golden

(Map 60)

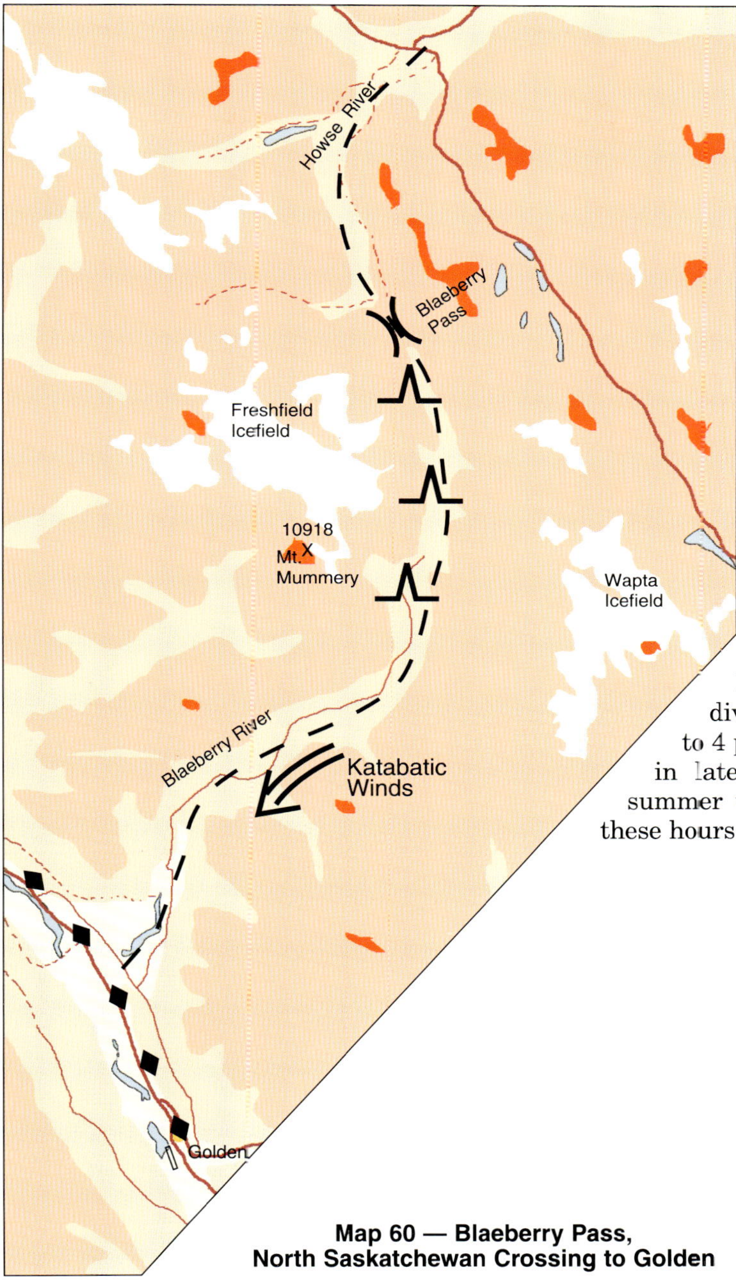

Map 60 — Blaeberry Pass, North Saskatchewan Crossing to Golden

Another common route into the Trench is Blaeberry Pass along the Howse and Blaeberry Rivers. This can be a very windy area due to the narrow valley. An extremely bad spot, even for helicopters, is at Mummery Creek. West of this location, winds and turbulence diminish rapidly except for extreme katabatic winds on sunny summer evenings. The differential solar heating of the forests and the glaciers tends to create chaotic wind patterns in summer months. On the nicest day pilots can get "beaten up" in the Lake Louise area and along the divide, mainly from 10 a.m. to 4 p.m. before it settles down in late afternoon. Pilots flying summer tours avoid flying during these hours.

Kootenay Crossing, Bow River to Radium

(See map 26, page 88)

Kamloops to Hope, Coquihalla

(See maps 35, north and south, pages 97 and 98)

Following the highway is generally feasible but low cloud can build up in the higher ground around Lac La Jeune. A good alternate is to use the Nicola Valley but be aware that you can run into low cloud over the hills just to the south of Kamloops. The route from the Toll Booth down to just south of the snowshed is subject to frequent low ceilings and visibility in snow during the winter. Conditions are described as being similar in characteristic to the weather around Hope but that it occurs over a longer run. Precipitation results in a very rapid lowering of ceilings and visibility. Snow showers can result in an instant lowering of the visibility to near zero.

Lee wave turbulence can be encountered near Merritt between 10,000 and 16,000 feet. Low level turbulence is usually confined to the area near Hope and along the route south of the Coquihalla Summit. When strong southerly winds aloft are occurring, expect moderate occasional severe turbulence at 6,000 to 7,000 feet near Yahk Mountain and Zopkios Ridge.

Kamloops to Vernon via Falkland

(See map 54 for details, page 114)

Considerable low cloud can be found in the narrow passes through the high ground. Occasional turbulence is reported near Falkland.

Salmon Arm to Osoyoos

(Salmon Arm to Vernon is detailed in map 54, page 114)

(Vernon to Osoyoos Map 61, opposite)

The base of the valley cloud that extends out of the Okanagan and across the Shuswap area is usually at the same height except for an area around Enderby/Armstrong and southern Kalamalka Lake where ceilings can be considerably lower.

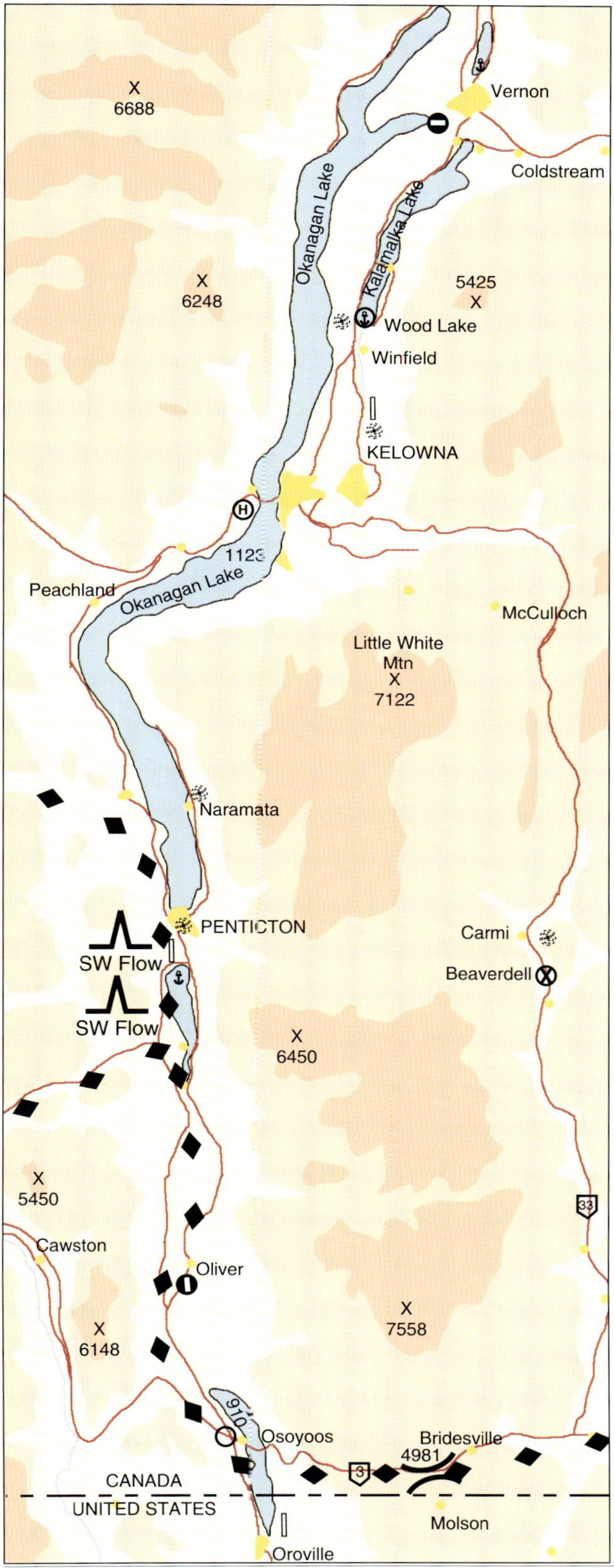

Penticton and Vicinity

(Map 61)

Penticton is a frequent stopover place for aircraft traveling across the interior. In general the weather is excellent except when major systems are moving through the area. Fog is rare and even the valley cloud that is so persistent in the winter usually remains between 3,000 and 4,000 feet. A unique hazard exists in this area when strong southwesterly flow aloft generates lee waves to the lee of the Coast Mountains. While not necessarily turbulent, these waves frequently generate a strong downflow of up to 1,400 feet per minute to the west of the airport.

Experienced pilots were emphatic with the following warning, "Do not try enroute climb out of Penticton. Rather climb south then turn to follow the VFR route westward."

Map 61 — Vernon to Osoyoos

Penticton to Princeton

(Map 62)

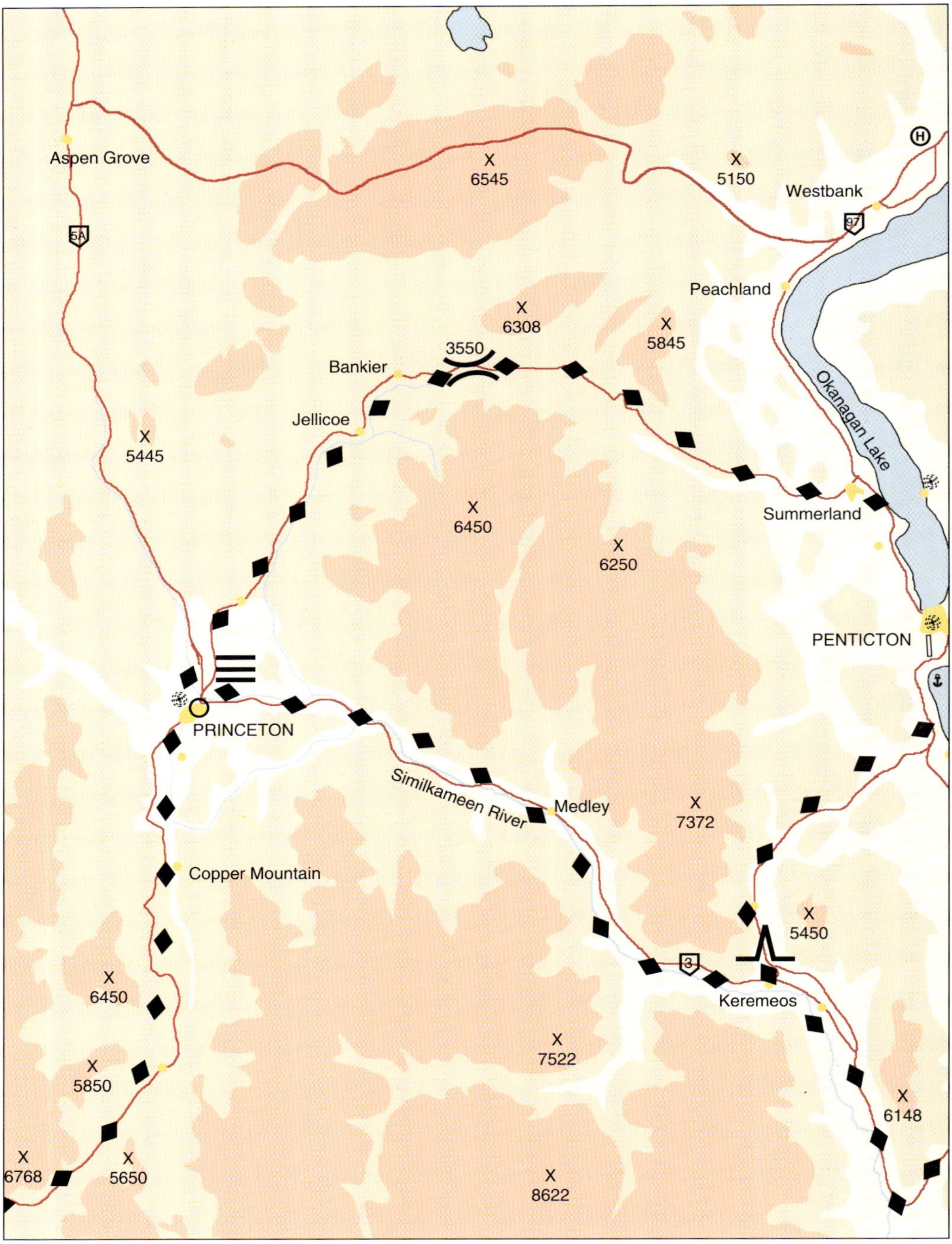

Map 62 — Penticton to Princeton

Vicinity Princeton

(Map 62, previous page)

Princeton lies within a bowl that is often filled with fog during the fall and spring. Fog is not so prevalent during the winter, but low valley-type cloud is often present. The approach into Princeton is quite narrow from the south and blowing snow coming off the ridges can at times reduce visibility significantly.

Osoyoos to Grand Forks

(Map 63)

Map 63 — Osoyoos to Grand Forks

Okanagan Lake

G. Fricska

Grand Forks to Castlegar

(Map 64)

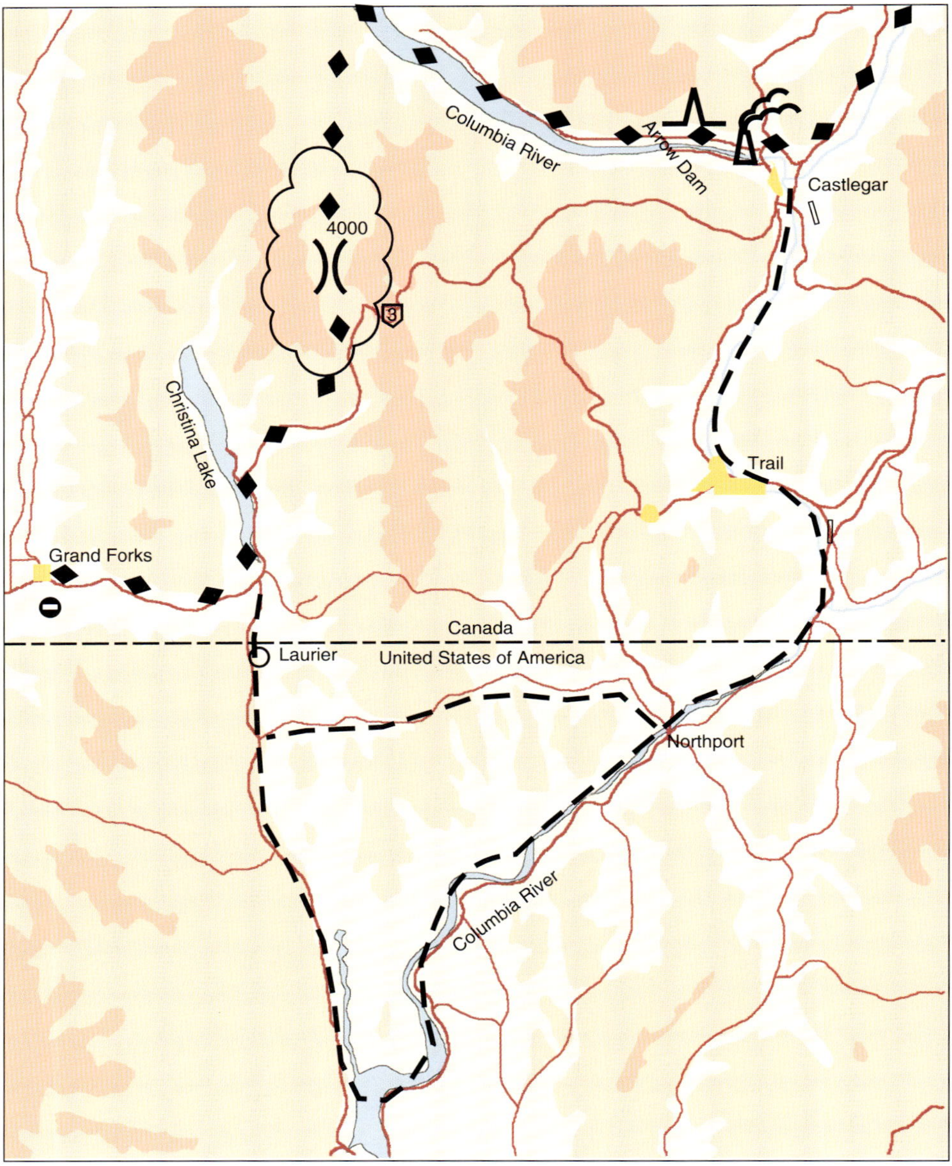

Map 64 — Grand Forks to Castlegar

Strong winds can produce turbulence near the Arrow Dam where the Columbia River Valley narrows. The pulp mill just west of Castlegar enhances low cloud and fog. Once west of the Arrow Dam, ceilings and visibility improve. The route from the Lower Arrow Lake to Christina Lake is very narrow and can be blocked by low cloud.

The Columbia River route to Northport (in U.S.) then west just south of the U.S. border then north to Christina Lake is much drier and lower than the route from Arrow Lake to Christina Lake.

Castlegar to Creston, Highway 3

(Map 65)

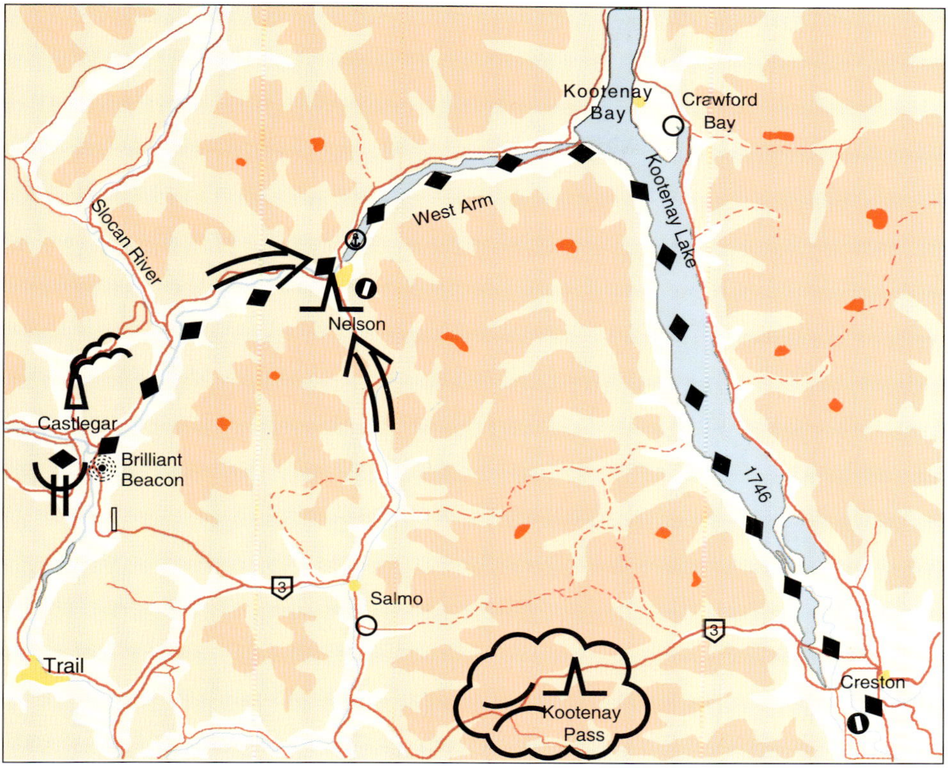

Map 65 — Castlegar to Creston via Highway 3

The high mountain pass between Salmo and Creston makes this route unpopular with most pilots. Weather along this route has forced a few emergency landings on the highway. In addition to low cloud, strong winds with wind shear are common.

Castlegar to Creston via Nelson and Kootenay Lake

(Map 65, above)

Low level winds can be strong in the Kootenay Bay area Converging winds where the valleys come together here and at Nelson are reported to be worse in a west to southwest flow. Nelson airport is known to experience crosswinds on a regular basis. In general, weather is reported to be better east of Nelson than between Nelson and Castlegar. In fact, pilots have reported some of the worst icing ever seen north of the Castlegar Airport, over the Brilliant Beacon. This icing is typically between 10,000 and 16,000 feet.

Crawford Bay to Cranbrook

(Map 66)

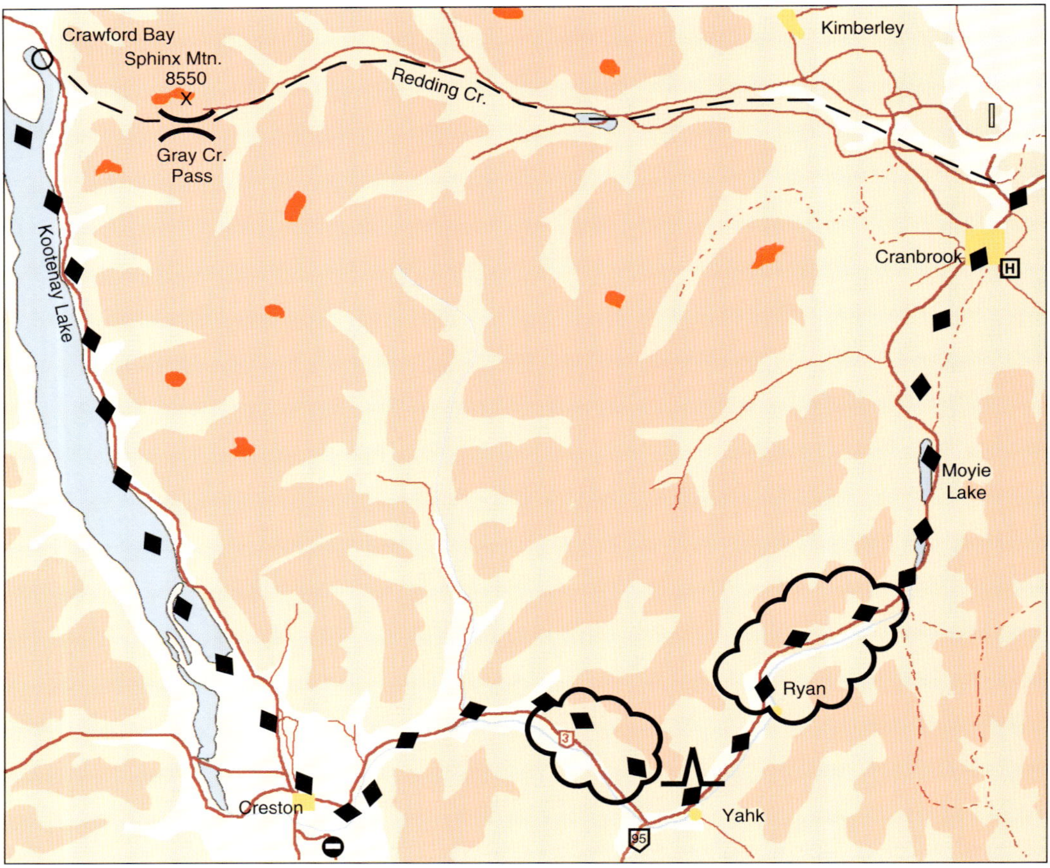

Map 66 — Crawford Bay to Cranbrook

This is the preferred route by many pilots traveling between Castlegar and Cranbrook. The highest point is Grey Creek Pass approximately 5 miles east of Crawford Bay. In the pass itself, it is recommended that you follow the road rather than the power line. When traveling west to east after crossing the pass, the weather tends to improve along the rest of the route. To get through the pass, ceilings must be higher than 7,500 feet ASL. If the route closes behind you there is no way out, you are committed to going east into the Trench. When traveling east to west, route finding is more difficult due to the many side valleys.

Creston to Cranbrook

(Map 66, above)

The Moyie River Valley between Yahk and Moyie Lake receives more precipitation than other areas during the summer. The corner near Yahk on occasion is turbulent and can distract pilots into taking the southern valley into the U.S. by mistake. There are two areas where low cloud tends to linger: between Moyie and Ryan and near the junctions of Hwy. 3 and 95 and extending 5 miles to the west.

Cranbrook Airport And Vicinity

(Map 67)

Cranbrook Airport tends to be representative of weather along the Trench. However, low cloud is more common over the Kootenay River. Although winds are not very strong, the prevailing westerly wind is across the runway. Shear has also been reported as significant within 200 feet above the runway. If the weather is bad at the airport, the Trench is more than likely closed.

Cranbrook to Crowsnest Pass

(Map 67)

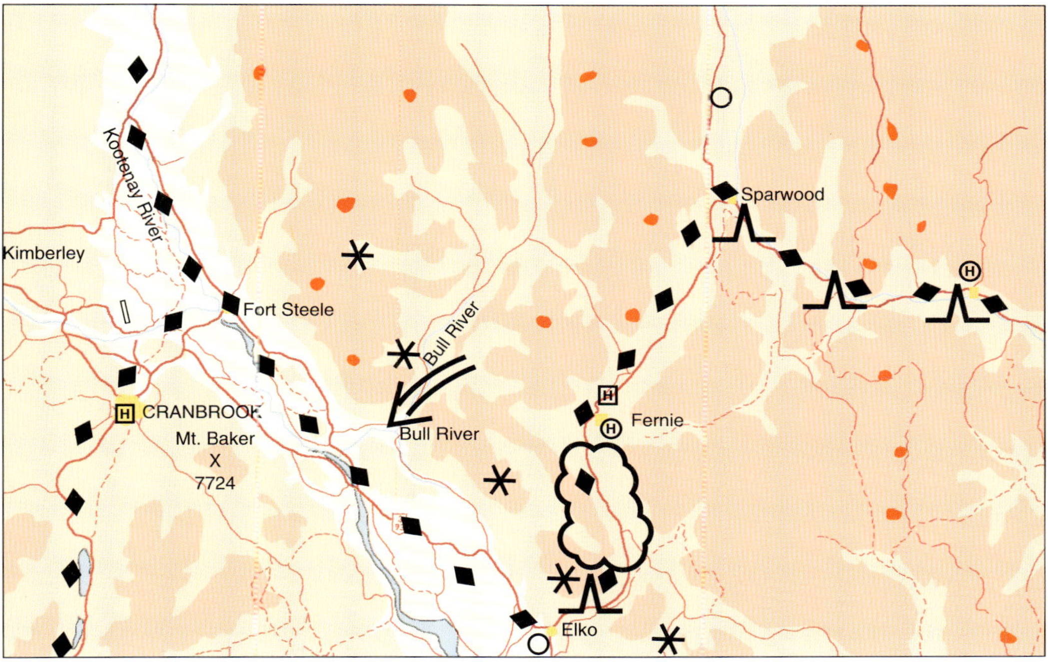

Map 67 — Cranbrook to Crowsnest Pass

The route from Sparwood to Blairmore (in Alberta) is well known for its strong winds (up to 80 knots) and turbulence. Planes have been forced down near Blairmore by severe turbulence. Local pilots say, “A nice day in Blairmore is like winning the lottery.” On most days the north side of the valley is less turbulent than the south side. Drier weather is experienced along the Iron Creek from Fernie to Bull River. The pass between Fernie and Bull River is easy to see from just west of Fernie. If this pass is closed then similar conditions likely exist along the highway to the south of Fernie. A heavy snow belt runs along the Rockies from the Bull River to the Flathead River. Strong outflow winds are common out of the Bull River valley.

> *Weather Hint:* The weather at Sparwood Airport is usually better than the weather at Fernie.

YUKON

The Yukon possesses terrain and associated aviation hazards which are more diverse than anywhere in North America. This manual will present only a brief overview of major weather hazards along the Alaska Highway. Specific weather hazards for remote areas of the Yukon are well beyond the scope of this manual. Experienced pilots are the only reliable sources of information. The Yukon Highways Department can provide weather information as they use radio communication to report conditions each morning.

The highest mountain ranges in North America lie to the southwest, between the Yukon and the Pacific Ocean. Given that most of the major weather systems originate in the Pacific, the Yukon lies in a rain shadow, with relatively dry weather. Unlike many routes in southern British Columbia, the valleys in the Yukon tend to be rather wide. Thus, funneled and channeled winds are not as severe. Lee waves are common when southwesterly winds aloft cross the mountains. These cannot be avoided in many cases; however, local pilots must be aware of their existence and ride the waves. Another major problem in summer months can be smoke due to forest fires. Smoke can close vast areas due to visibilities well below VFR criteria.

Alaska Highway Watson Lake to Teslin

(Map 68 and Map 69)

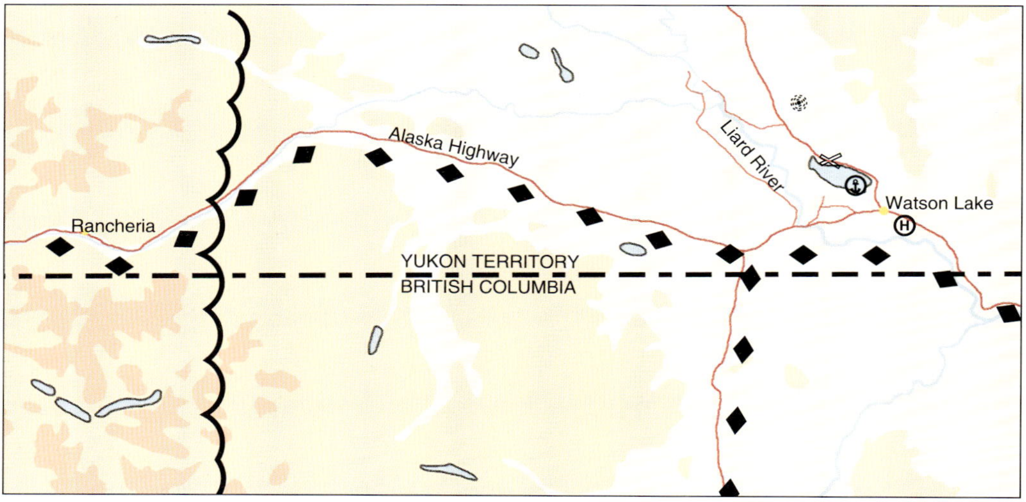

Map 68 — Alaska Highway, Watson Lake to Teslin

The relatively high route between Watson Lake and Teslin is well known for its propensity for low cloud and fog. The elevation between Rancheria and Swift River is more than 1,000 feet greater than Watson Lake or Teslin with considerably higher mountains on both sides of the valley. Even when both Watson Lake and Teslin report good VFR weather, low cloud can often close much of the route. This same area is also prone to heavy snow, even in late summer or early autumn. As with many deeper lakes, low convective-type cloud can be a problem over Teslin and Watson Lakes in the autumn until freeze up.

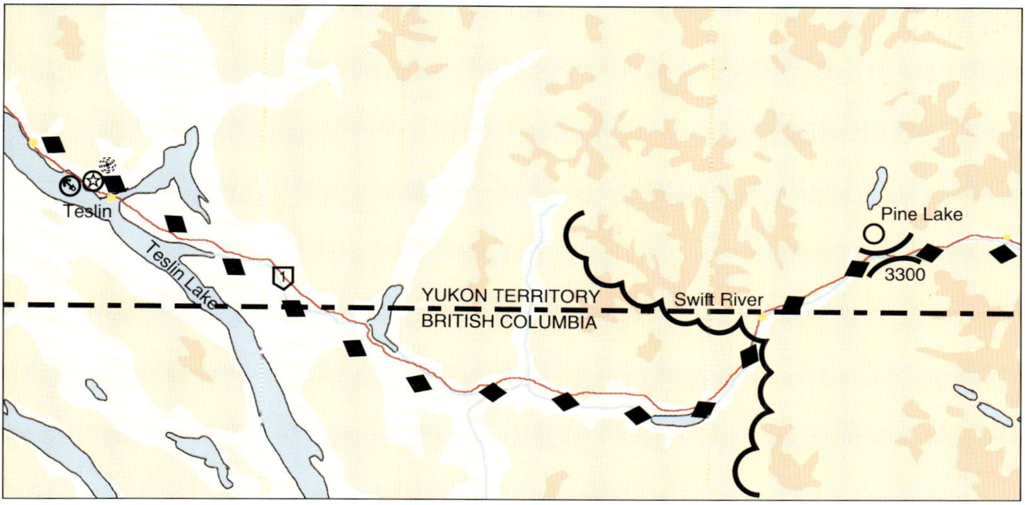

Map 69 — Alaska Highway, Watson Lake to Teslin

Alaska Highway Teslin to Whitehorse

(Not shown)

Low cloud and fog are less common west and north of Teslin. The route is known for its generally good weather. As with most of the Yukon, turbulence can be strong when winds blow across the north-south mountain ranges.

Whitehorse Region

(Map 70)

The Whitehorse Airport weather station is known to be a relatively good indicator of weather in the area. Southerly winds can be strong although they do not tend to produce significant turbulence. On the other hand, winds blowing from the southwest can produce moderate or greater turbulence, particularly to the lee of Pilot and Flat Mountains, to the north of the airport. This turbulence is usually lighter further east, over Lake Laberge.

Lake Laberge can produce considerable low cloud in the autumn, once temperatures begin to fall below freezing. This cloud can, at times, drift southward over the Whitehorse Airport.

Other lakes to the south and southeast can also produce low cloud and fog in late summer and autumn.

Summer convective clouds tend to develop almost every day over Joe Mountain, to the northeast of the city. In most other cases, bad weather at Whitehorse is indicative of bad or worse weather along all surrounding routes.

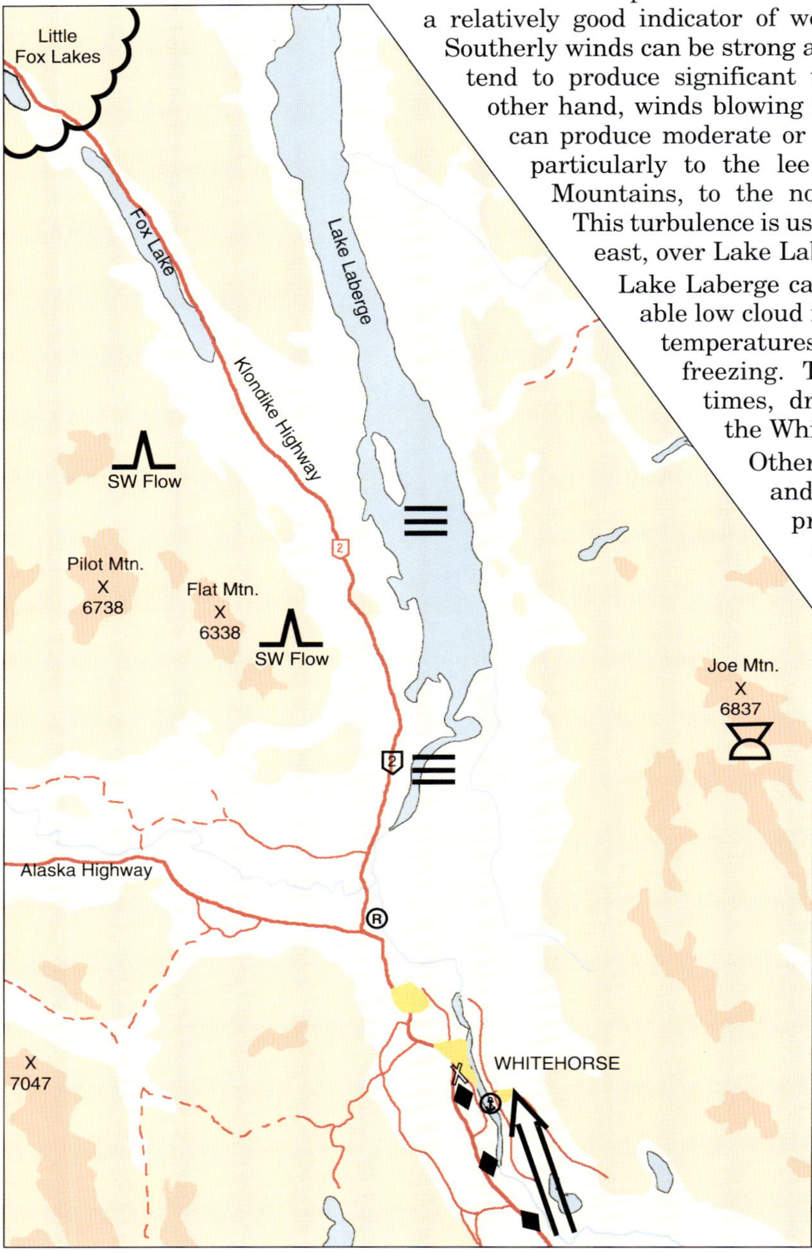

Map 70 — Whitehorse region

Alaska Highway Whitehorse to Haines Junction

(Map 71)

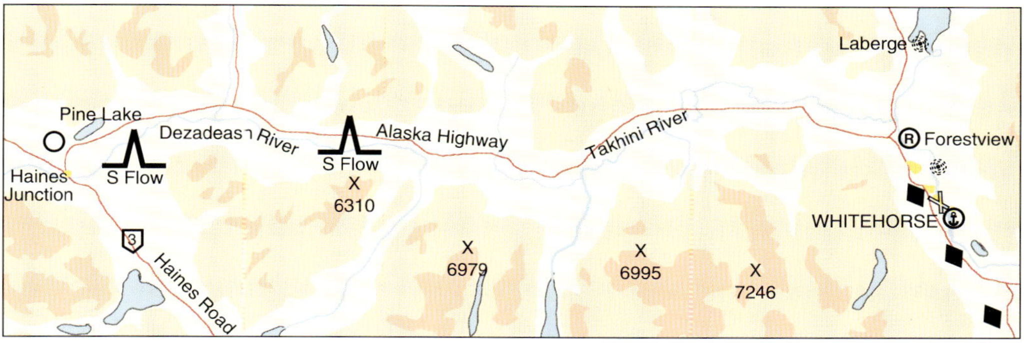

Map 71 — Alaska Highway Whitehorse to Haines Junction

The route from Whitehorse to Haines Junction follows the very wide and relatively low Takhini and Dezadeash River Valleys. This route is reported to experience few aviation weather hazards with the exception of significant turbulence when southerly winds blow across the valleys. On windy days with southerly winds, pilots describe a wild ride through Haines Junction due to winds blowing over the high mountains to the south.

Alaska Highway Haines Junction to Burwash

(Map 72)

Just out of Haines Junction, this route crosses what typically is the worst weather on the route.

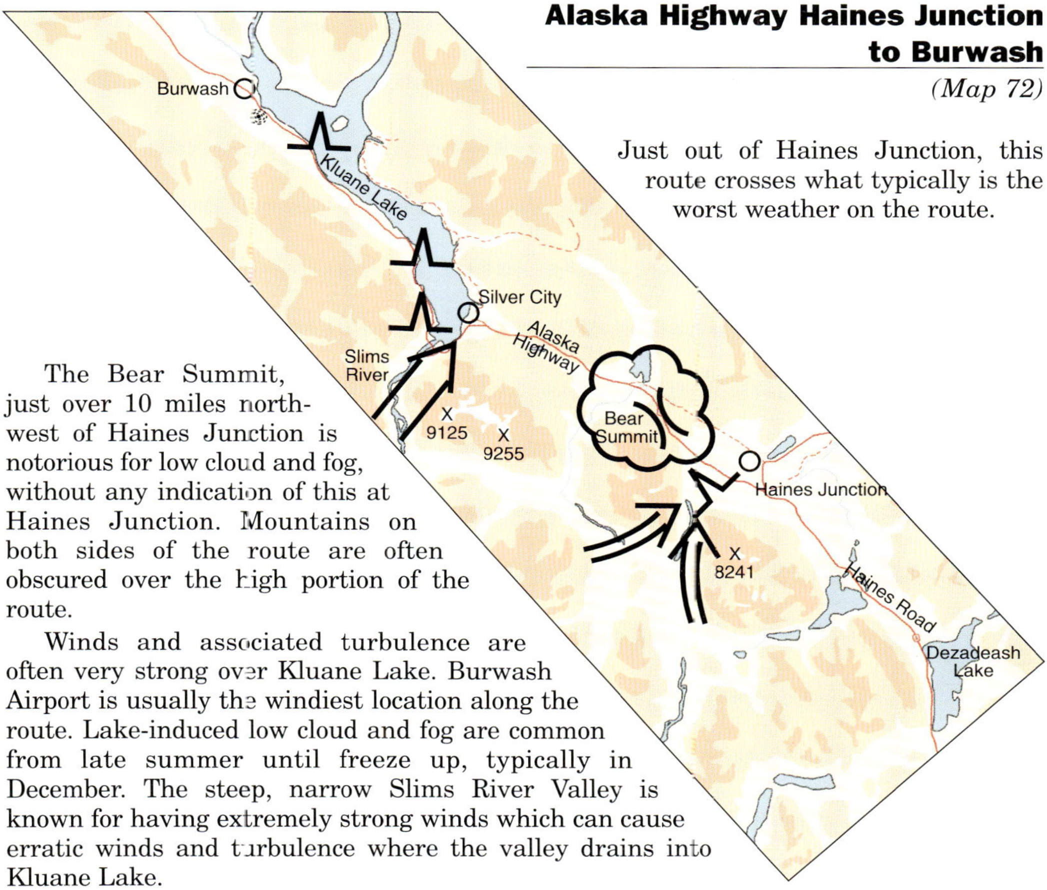

The Bear Summit, just over 10 miles northwest of Haines Junction is notorious for low cloud and fog, without any indication of this at Haines Junction. Mountains on both sides of the route are often obscured over the high portion of the route.

Winds and associated turbulence are often very strong over Kluane Lake. Burwash Airport is usually the windiest location along the route. Lake-induced low cloud and fog are common from late summer until freeze up, typically in December. The steep, narrow Slims River Valley is known for having extremely strong winds which can cause erratic winds and turbulence where the valley drains into Kluane Lake.

Map 72 — Alaska Highway Haines Junction to Burwash

Alaska Highway Burwash to the Alaska Border

(Not shown)

This portion of the route lies in one of the driest areas of the Yukon.

Poor aviation weather tends to occur, on occasion, with the northwesterly flows. In this case, low cloud can push all the way to Burwash or even Haines Junction.

Klondike Highway Whitehorse to Dawson

(Not shown)

Low cloud and fog tend to linger over the region near Little Fox Lakes, approximately halfway between Whitehorse and Carmacks (shown on Whitehorse and vicinity, map 70, page 130). From late August until October, morning fog tends to close the Dawson airport with the fog extending for 10 to 15 miles down both the Yukon and Klondike Rivers.

Haines and Skagway Roads

(Not shown)

VFR conditions along these spectacular routes are, unfortunately, rare. Low cloud, strong winds and significant icing occur along the coast, although weather along the route may appear perfect for most of the journey.

Dempster Highway: Klondike Highway to Arctic Red River

(Not shown)

The bad weather along this route tends to always occur over the Ogilvie and Richardson Mountains with low cloud and turbulence which often reaches severe. Violent winds can occur over the Richardson Mountains although most of the weather reportedly occurs below 7,000 feet ASL.

GLOSSARY OF WEATHER TERMS

anabatic wind - a local wind which blows up a slope heated by sunshine.

advection - the horizontal transportation of air or atmospheric properties.

air density - the mass density of air expressed as weight per unit volume.

air mass - an extensive body of air with uniform conditions of moisture and temperature in the horizontal.

anticyclone - an area of high atmospheric pressure which has a closed circulation that is anticyclonic (clockwise) in the Northern Hemisphere.

cat's paw - ripple signature on water given by a strong downdraft below a towering cumulus or cumulonimbus cloud. The downdraft makes a cat paw-like imprint on the water. A good indication of turbulence and wind shear.

ceiling - either (a) the height above the surface of the base of the lowest layer of clouds or obscuring phenomena (i.e. smoke) that hides more than half of the sky; (b) the vertical visibility into an obstruction to vision (i.e. fog).

chinook - a warm dry wind blowing down the eastern slopes of the Rocky Mountains and over the adjacent plains.

clear air turbulence (CAT) - turbulence encountered in air where no clouds are present; normally associated with high level turbulence.

clear icing - generally, the formation of a layer or mass of ice which is relatively transparent because of its homogeneous structure and smaller number and size of air spaces; synonymous with glaze.

climate - the statistical collection of long-term (usually decades) weather conditions at a point; may be expressed in a variety of ways.

cold front - the leading edge of an advancing cold air mass.

convection - atmospheric motions that are predominately vertical, resulting in the vertical transport and mixing of atmospheric properties.

convergence - a condition that exists when the distribution of winds in a given area is such that there is a net horizontal inflow of air into the area; the effect is to create lift.

cumuliform - a term descriptive of all convective clouds exhibiting vertical development.

cyclone - an area of low atmospheric pressure which has a closed circulation that is cyclonic (counterclockwise) in the Northern Hemisphere.

deepening - a decrease in the central pressure of a pressure system; usually applied to a low.

deformation zone - the region to the northeast of a low pressure system where the air parcels (and clouds) undergo stretching along the east to west axis and contraction along the north to south axis. Meteorogically, this is an area of significant cloud amounts and precipitation due to the induced vertical ascent *(see fig. G1)*.

disturbance - applied loosely: (a) any small-sized low pressure system; (b) an area where the weather, wind, pressure, etc. shows signs of cyclonic development; (c)

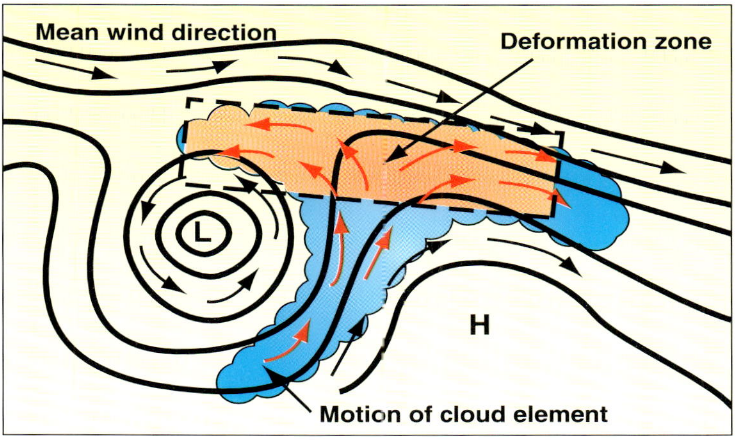

Fig. G1 — Deformation zone. Moisture (cloud) moving up from the south ahead of the low is stretched out or deformed by the changing wind direction in the *deformation zone*

any deviation in flow or pressure that is associated with a disturbed state in the weather; and (d) any individual circulatory system within the primary circulation of the atmosphere.

divergence - a condition that exists when the distribution of winds in a given area is such that there is a net horizontal outflow of air from the area; effect is to create subsidence.

downdraft - a small scale downward current of air; observed on the lee side of large objects that restrict the smooth flow of air or in or near precipitation areas associated with cumuliform clouds.

downburst - an exceptionally strong downdraft beneath a thunderstorm.

filling - an increase in the central pressure of a pressure system; applied to a low.

front - a surface, interface or transition zone of discontinuity between two adjacent air masses of different densities.

funnel cloud - a tornado cloud or vortex cloud extending downward from the parent cloud but not reaching the ground.

gust - a sudden, rapid and brief increase in wind speed. In Canada gusts are reported when the highest peak speed is at least 5 knots higher than the average wind and the highest peak speed is at least 15 knots.

gust front - the leading edge of the downdraft outflow ahead of a thunderstorm.

high - an area of high barometric pressure; a high pressure system.

icing - in general, any deposit of ice forming on an object.

instability - a state of the atmosphere such that the vertical distribution of temperature is such that a parcel displaced from its initial position will continue to ascend.

inversion - an increase of temperature with height - a reversal of the normal decrease of temperature with height.

isothermal layer - equal or constant temperature with height.

jet stream - a quasi-horizontal stream of wind concentrated within a narrow band; generally located just below the tropopause.

katabatic wind - a downslope wind.

knot - a unit of speed equal to one nautical mile per hour.

lapse rate - the rate of change of an atmospheric variable (usually temperature) with height.

lee wave - any stationary wave disturbance caused by a barrier in a fluid flow; also called mountain wave or standing wave.

lightning - generally, any and all forms of visible electrical discharge produced by a thunderstorm.

low - an area of low barometric pressure; a low pressure system.

meteorology - the science of the atmosphere.

microburst - a narrow band of extremely violent winds embedded within a downburst; a thin ribbon of wind about a mile across that lasts for only a few minutes but is capable of forcing aircraft into the ground.

mixed icing - the formation of a white or milky and opaque layer of ice that demonstrates an appearance that is a composite of rime and clear icing.

occluded front - a front that is no longer in contact with the surface.

orographic - of, pertaining to, or caused by forced uplift of air over high ground.

outflow - generally, a condition where air is flowing from the interior land area through mountain passes, valleys and inlets onto the coastal areas; used most commonly in winter when cold Arctic air spreads onto the coastal area and adjoining sea.

overrunning - generally, a condition when warm air overtakes or is lifted by colder denser air.

parcel - a small volume of air, small enough to contain uniform distribution of meteorological properties, and large enough to remain relatively self-contained and respond to all meteorological processes.

plow wind - usually associated with the spreading out of a downburst from a thunderstorm; a strong, straight-line wind in advance of a thunderstorm that often results in severe damage.

precipitation - any and all forms of water particles, whether liquid or solid, that fall from the atmosphere and reach the surface.

quasi-stationary front - a front that is stationary or nearly so; commonly called stationary front.

ridge - an elongated area of relatively high atmospheric pressure; also called ridge line.

rime icing - the formation of a white or milky and opaque granular deposit of ice formed by the rapid freezing of supercooled water droplets.

saturation - the condition in the atmosphere where actual water vapour present is the maximum possible at the existing temperature.

shower - precipitation from cumuliform cloud; characterized by suddenness of beginning and ending, by rapid changes in intensity, and usually by rapid change in the appearance of the sky.

squall - essentially gusts of longer duration. In Canada a squall is reported when the wind increases by at least 15 knots over the average speed for a duration of at least 2 minutes and the wind reaches a speed of at least 20 knots.

squall line - a non-frontal line or narrow band of active thunderstorms.

stability - a state of the atmosphere where the vertical distribution of temperature is such that a parcel will resist displacement from its initial position.

stratiform - term descriptive of clouds of extensive horizontal development; flat, lacking definition.

stratosphere - the atmospheric layer above the tropopause; characterized by slight increase in temperature from base to top, very stable, low moisture content and absence of cloud.

subsidence - the downward motion of air over a large area resulting in dynamic heating.

supercooled water - liquid water at temperatures below freezing.

thunderstorm - a local storm invariably produced by a cumulonimbus cloud, and always accompanied by lightning and thunder.

tornado - a violently rotating column of air, pendant from a cumulonimbus cloud, and nearly always observed as "funnel-shaped;" other names are cyclone and twister.

tropopause - the transition zone between the troposphere and the stratosphere; characterized by an abrupt change in lapse rate.

troposphere - the portion of the earth's atmosphere from the surface to the tropopause; characterized by decreasing temperature with height and appreciable water vapour. Often referred to as the weather layer.

trough - an elongated area of relatively low atmospheric pressure; also called trough line.

trowal - a trough of warm air aloft; related to occluded front.

turbulence - any irregular or disturbed flow in the atmosphere.

updraft - a localized upward current of air.

upper front - any frontal zone which is not manifested at the surface.

virga - water or ice particles falling from a cloud, usually in wisps or streaks, and evaporating completely before reaching the ground.

warm front - the trailing edge of retreating cold air.

weather - the instantaneous conditions or short term changes at a point; as opposed to climate.

wind - air in motion relative to the earth's surface; normally horizontal motion.

wind direction - the direction from which the wind is blowing.

wind speed - rate of wind movement expressed as distance per unit time.

wind shear - the rate of change of wind direction and/or speed per unit distance; conventionally expressed as vertical and horizontal wind shear.

zonal wind - a west wind; conventionally used to describe large-scale flow that is neither cyclonic or anticyclonic; also called zonal flow.

If you have any comments or information that you feel might add to future versions of the Aviation Weather Hazards Manual, please mail to the address below.

Name: ______________________

Address: ______________________

____

____

Overall comments:

____

____

____

____

Additional local weather hazards:

____

____

____

____

____

____

For more information on purchasing additonal copies,
please write to:
Aviation Weather Hazards Manual
3140 College Way
Kelowna, B.C.
V1V 1V9
or e-mail to kent.johnson@ec.gc.ca or
john.mullock@ec.gc.ca